APPLIED RESEARCH ON POLYMER COMPOSITES

AAP Research Notes on
Polymer Engineering Science and Technology

APPLIED RESEARCH ON POLYMER COMPOSITES

Edited by
**Pooria Pasbakhsh, PhD, A. K. Haghi, PhD, and
Gennady E. Zaikov, DSc**

Apple Academic Press Inc. | Apple Academic Press Inc.
3333 Mistwell Crescent | 9 Spinnaker Way
Oakville, ON L6L 0A2 | Waretown, NJ 08758
Canada | USA

© 2015 by Apple Academic Press, Inc.

First issued in paperback 2021

Exclusive worldwide distribution by CRC Press, a member of Taylor & Francis Group
No claim to original U.S. Government works

ISBN 13: 978-1-77463-354-0 (pbk)
ISBN 13: 978-1-77188-038-1 (hbk)

Library and Archives Canada Cataloguing in Publication

Applied research on polymer composites/edited by Pooria Pasbakhsh, PhD, A.K. Haghi, PhD, and Gennady E. Zaikov, DSc.

(AAP research notes on polymer engineering science and technology series)
Includes bibliographical references and index.
ISBN 978-1-77188-038-1 (bound)
1. Polymeric composites. 2. Polymers. I. Pasbakhsh, Pooria, editor II. Haghi, A. K., editor III. Zaikov, G. E. (Gennadi™i Efremovich), 1935-, author, editor IV. Series: AAP research notes on polymer engineering science and technology series

TA418.9.C6A66 2015   620.1'92   C2015-900211-7

Library of Congress Cataloging-in-Publication Data

Applied research on polymer composites / editors, Pooria Pasbakhsh, A.K. Haghi, and Gennady E. Zaikov.

pages ; cm
Includes bibliographical references and index.
ISBN 978-1-77188-038-1 (hardcover : alk. paper)
1. Polymeric composites. 2. Polymers. I. Pasbakhsh, Pooria, editor. II. Haghi, A. K., editor. III. Zaikov, G. E. (Gennadii Efremovich), 1935- , editor.
[DNLM: 1. Polymers. QD 381]

TA455.P58A685 2015   620.1'92--dc23   2014049942

Apple Academic Press also publishes its books in a variety of electronic formats. Some content that appears in print may not be available in electronic format. For information about Apple Academic Press products, visit our website at **www.appleacademicpress.com** and the CRC Press website at **www.crcpress.com**

ABOUT THE EDITORS

Pooria Pasbakhsh, PhD

Dr Pooria Pasbakhsh's research has been centered on the development and application of computational approaches and experimental tools on modern polymers, halloysite nanotubes, and polymer clay nanocompoites. He is currently Lecturer at School of Engineering, Monash University Sunway Campus, Malaysia.

Gennady E. Zaikov, DSc

Gennady E. Zaikov, DSc, is Head of the Polymer Division at the N. M. Emanuel Institute of Biochemical Physics, Russian Academy of Sciences, Moscow, Russia, and Professor at Moscow State Academy of Fine Chemical Technology, Russia, as well as Professor at Kazan National Research Technological University, Kazan, Russia. He is also a prolific author, researcher, and lecturer. He has received several awards for his work, including the Russian Federation Scholarship for Outstanding Scientists. He has been a member of many professional organizations and on the editorial boards of many international science journals.

A. K. Haghi, PhD

A.K. Haghi, PhD, holds a BSc in urban and environmental engineering from the University of North Carolina (USA); an MSc in mechanical engineering from North Carolina A&T State University (USA); a DEA in applied mechanics, acoustics and materials from Université de Technologie de Compiègne (France); and a PhD in engineering sciences from Université de Franche-Comté (France). He is the author and editor of 150 books and 1000 published papers in various journals and conference proceedings. Dr Haghi has received several grants, consulted for a number of major corporations, and is a frequent speaker to national and international audiences. Since 1983, he served as a professor at several universities. He is currently Editor-in-Chief of the *International Journal of Chemoinfor-*

matics and Chemical Engineering and *Polymers Research Journal* and on the editorial boards of many international journals. He is a member of the Canadian Research and Development Center of Sciences and Cultures (CRDCSC), Montreal, Quebec, Canada.

ABOUT AAP RESEARCH NOTES ON POLYMER ENGINEERING SCIENCE AND TECHNOLOGY

The AAP Research Notes on Polymer Engineering Science and Technology reports on research development in different fields for academic institutes and industrial sectors interested in polymer engineering science and technology. The main objective of this series is to report research progress in this rapidly growing field.

BOOKS IN THE AAP RESEARCH NOTES ON POLYMER ENGINEERING SCIENCE AND TECHNOLOGY SERIES

- **Functional Polymer Blends and Nanocomposites: A Practical Engineering Approach**
  Editors: Gennady E. Zaikov, DSc, Liliya I. Bazylak, PhD, and A. K. Haghi, PhD

- **Polymer Surfaces and Interfaces: Acid-Base Interactions and Adhesion in Polymer-Metal Systems**
  Irina A. Starostina, DSc, Oleg V. Stoyanov, DSc, and Rustam Ya. Deberdeev, DSc

- **Key Technologies in Polymer Chemistry**
  Editors: Nikolay D. Morozkin, DSc, Vadim P. Zakharov, DSc, and Gennady E. Zaikov, DSc

- **Polymers and Polymeric Composites: Properties, Optimization, and Applications**
  Editors: Liliya I. Bazylak, PhD, Gennady E. Zaikov, DSc, and A. K. Haghi, PhD

- **Applied Research on Polymer Composites**
  Editors: Pooria Pasbakhsh, PhD, A. K. Haghi, PhD, and Gennady E. Zaikov, DSc

- **High-Performance Polymers for Engineering-Based Composites**
  Editors: Omari V. Mukbaniani, DSc, Marc J. M. Abadie, DSc, and Tamara Tatrishvili, PhD

Dedicated to

Professor Gennady E. Zaikov on the Occasion of
His Eightieth Birthday, 7 January 2015

CONTENTS

LIST OF CONTRIBUTORS

L. F. Akhmetshina
OJSC "Izhevsk Electromechanical Plant – Kupol"

M. P. Anachkov
S. K. Rakovsky, Institute of Catalysis, Bulgarian Academy of Sciences, Acad. G. Bonchev bl.11, 1113 Sofia, Bulgaria, Email: rakovsky@ic.bas.bg

M. I. Artsis
Institute of Biochemical Physics after N. M. Emanuel, Kosygina 4, Moscow 119991, Russia, Email chembio@sky.chph.ras.ru
M. MehdiPour, University of Guilan, Rasht, Iran

V. A. Babkin
Volgograd State University Architecture and Civil Engineering Sebrykov Department, Email: Babkin_v.a@mail.ru.

D. V. Bagrov
Faculty of Biology, Moscow State University, Leninskie gory, Moscow 119992, Russia

A. A. Berlin
N. N. Semenov's Institute of Chemical Physics, RAS, Moscow, 119996 RF

A. P. Bonartsev
A. N. Bach's Institute of Biochemistry, Russian Academy of Sciences, Leninskiy prosp, 119071 Moscow, Russia and Faculty of Biology, Moscow State University, Leninskie gory, 119992 Moscow, Russia

G. A. Bonartseva
A. N. Bach's Institute of Biochemistry, Russian Academy of Sciences, Leninskiy prosp, Moscow 119071, Russia

A. P. Boskhomodgiev
A. N. Bach's Institute of Biochemistry, Russian Academy of Sciences, Leninskiy prosp, 119071 Moscow, Russia

A. A. Burkov
Vyatka State University, Kirov, 610000 RF, Email: rubber_zerg@mail.ru

M. A. Chashkin
OJSC "Izhevsk Electromechanical Plant – Kupol"

E. A. Filatova
A. N. Bach's Institute of Biochemistry, Russian Academy of Sciences, Leninskiy prosp, 119071 Moscow, Russia

S. V. Fomin
Vyatka State University, Kirov, 610000 RF, Email: rubber_zerg@mail.ru

Zenon Foltynowicz
Faculty of Commodity Science, Poznan University of Economics, 61-875 Poznań, Poland

K. Z. Gumargalieva
N. N.Semenov's Institute of Chemical Physics, RAS, Moscow, 119996 RF

A. L. Iordanskii
A. N. Bach's Institute of Biochemistry, Russian Academy of Sciences, Leninskiy prosp, 119071 Moscow, Russia and N.N.Semenov Institute of Chemical Physics, Russian Academy of Sciences, Moscow 119991, Russia

E. A. Ivanov
A. N. Bach's Institute of Biochemistry, Russian Academy of Sciences, Leninskiy prosp, 119071 Moscow, Russia

N. V. Khokhriakov
Basic Research – High Educational Centre of Chemical Physics and Mesoscopy, Udmurt Scientific Centre, Ural Division, Russian Academy of Sciences, Izhevsk and Izhevsk State Agricultural Academy

V. I. Kodolov
M. T. Kalashnikov Izhevsk State Technical University and Basic Research – High Educational Centre of Chemical Physics and Mesoscopy, Udmurt Scientific Centre, Ural Division, Russian Academy of Sciences, Izhevsk

G. V. Kozlov
Institute of Applied Mechanics of Russian Academy of Sciences, Leninskii pr, Moscow 119991, Russian Federation, Email: IAM@ipsun.ras.ru

Hieronim Maciejewski
Faculty of Chemistry, Adam Mickiewicz University, 61-614 Poznań, Poland and Poznań Science and Technology Park of Adam Mickiewicz University Foundation, 61-612 Poznań, Poland

T. K. Makhina
A. N. Bach's Institute of Biochemistry, Russian Academy of Sciences, Leninskiy prosp, 119071 Moscow, Russia

V. L. Myshkina
A. N. Bach's Institute of Biochemistry, Russian Academy of Sciences, Leninskiy prosp, Moscow 119071, Russia

Yu. N. Pankova
N. N. Semenov's Institute of Chemical Physics, RAS, Moscow, 119996 RF Email: aljordan08@gmail.com

Przemysław Pietras
Faculty of Chemistry, Adam Mickiewicz University, Poznań, Poland
Ryszard Fiedorow, Faculty of Chemistry, Adam Mickiewicz University, 61-614 Poznań, Poland

Ya. A. Polyotov
M. T. Kalashnikov Izhevsk State Technical University and Basic Research – High Educational Centre of Chemical Physics and Mesoscopy, Udmurt Scientific Centre, Ural Division, Russian Academy of Sciences, Izhevsk

O. A. Ponomarev
Pushino, Institute of Mathematical Problem of Biology Russian Akademy of Science, Email: olegpon36@mail.ru

A. I. Rakhimov
Volgograd State University, Email: organic@vstu.ru

N. A. Rakhimov
Volgograd State University, Email: organic@vstu.ru

A. V. Rebrov
A. V. Topchiev Institute of Petroleum Chemistry. Leninskiy prosp, 119071 Moscow, Russia

S. Z. Rogovina
N. N. Semenov's Institute of Chemical Physics, RAS, Moscow, 119996 RF

E. S. Titova
Volgograd State Technical University, Email: organic@vstu.ru
N. M. Emanuel Institute of Biochemical Physics of Russian Academy of Sciences, Moscow 119334, Russian Federation, Email: Chembio@sky.chph.ras.ru

V. V. Trineeva
Basic Research – High Educational Centre of Chemical Physics and Mesoscopy, Udmurt Scientific Centre, Ural Division, Russian Academy of Sciences, Izhevsk and Institute of Mechanics, Ural Division, Russian Academy of Sciences, Izhevsk

A. V. Vakhrushev
A. M. Lipanov, Institute of Mechanics, Ural Branch of the Russian Academy of Sciences, Izhevsk, Russia Email: postmaster@ntm.udm.ru

S. A. Yakovlev
A. N. Bach's Institute of Biochemistry, Russian Academy of Sciences, Leninskiy prosp, 119071 Moscow, Russia

Yu. G. Yanovskii
Institute of Applied Mechanics of Russian Academy of Sciences, Leninskii pra, Moscow 119991, Russian Federation, Email: IAM@ipsun.ras.ru

G. E. Zaikov
N. M. Emanuel Institute of Biochemical Physics, Russian Academy of Sciences, Moscow 119334 Russia, Email: chembio@sky.chph.ras.ru

LIST OF ABBREVIATIONS

AC	activated carbon
ACFs	activated carbons fibers
ACNF	activated carbon nanofiber
AFM	atomic force microscopy
AKDs	ketene dimers
AMF	advanced method of surface investigation
ASAs	alkenyl succinic anhydrides
ASM	atomicpower microscopy
BET	Brunauer–Emmett–Teller
BJH	Barrett–Joyner–Halenda
CNF	carbon nanofiber
DFT	density functional theory
DOM	dissolved natural organic matter
EDLC	electric double-layer capacitor
FGG	food grade gelation
FTIR	Fourier transform infrared spectroscopy
GCMC	Grand Canonical Monte Carlo
HK	Horvath–Kawazoe
LC	liquid chromatography
LIBs	lithium-ion batteries
MD	molecular dynamics
MFC	microbial fuel cell
MFCs	microbial fuel cells
MSC	molecular sieving carbon
MW	molecular weight
MWCNTs	multiwalled carbon nanotubes
PALS	positron annihilation lifetime spectroscopy
PBI	polybenzimidazol
PGAs	polyglycolides
PHAs	polyhydroxyalkanoates
PHB	poly(R)-3-hydroxybutyrate
PIB	polyisobutylene

PIs	polyimides
PLAs	polylactides
POs	primary ozonides
PSA	pressure swing adsorption
PSD	pore size distribution
PVA	poly (vinyl alcohol)
PVDF	poly (vinylidene fluoride)
RGG	research grade gelatin
RVC	reticulated vitreous carbon
SANS	small-angle neutron scattering
SEM	scanning electron microscopy
SOCs	synthetic organic chemicals
SPIP	scanning probe image processor
STM	scanning tunneling microscopy
SWCNTs	single-walled CNT
T&O	taste and odor
TEOS	tetraethoxysilan
TSA	temperature swing adsorption
WAXS	wide angle X-ray scattering
XRD	X-ray diffraction

LIST OF SYMBOLS

φ_{if} and φ_n	relative volume fractions
v_m	polymer matrix Poisson's ratio
ρ_n	density
ρ_n	nanocomposite density
ρ_n	nanofiller particle
φ_n	nanofiller volume fraction
v_{TC}	nanofiller (technical carbon)
a	lower linear scale of fractal behavior
c	nanoparticles concentration
C_∞	characteristic ratio
k	Boltzmann constant
k_d	effective rate constant of hydrolytic depolymerisation
k_h	effective hydrolysis constant
k_n	proportionality coefficient
K_T	isothermal modulus of dilatation
l_0	main chain skeletal bond length
M_m	constant molecular mass of monomer
P_n^0	initial number-average degree of polymerization at time 0
S_i	quadrate area
S_u	nanoshungite particles specific surface
T	testing temperature
W_n	nanofiller mass content

PREFACE

Polymers play a significant part in humans' existence. They have a role in every aspect of modern life, such as health care, food, information technology, transportation, energy industries, and so on. The speed of developments within the polymer sector is phenomenal and, at same time, crucial to meet the demands of today's and future life. Specific applications for polymers range from adhesives, coatings, painting, foams and packaging to structural materials, composites, textiles, electronic and optical devices, biomaterials, and many other uses in industries and daily life. Polymers are the basis of natural and synthetic materials. They are macromolecules and, in nature, are the raw material for proteins and nucleic acids, which are essential for human bodies.

Cellulose, wool, natural and synthetic rubber, plastics are all well-known examples of natural and synthetic types. Natural and synthetic polymers play a massive role in everyday life, and a life without polymers really does not exist. A correct understanding of polymers did not exist until the 1920s. In 1922, Staudinger published his idea that polymers were long chain molecules with normal chemical bonds holding them together. But for nearly 10 years, this idea did not attract much attention. Around this period, other researchers such as Carothers who tended toward Staudinger's idea discovered a type of synthetic material that could be produced by its constituent monomers. Later on, it was shown that as well as addition reaction, polymers could be prepared through condensation mechanism.

The introduction of plastics is associated with the twentieth century, but the first plastic material celluloid, was made in 1865. During the 1970s, clothes of polyester became fashionable, but by the 1980s, synthetics lost the popularity in favor of natural materials. Although people were less enthusiastic about synthetic fabrics for everyday wear, Gore-Tex and other synthetics became popular for outdoor and workout clothing. At the same time as the use of synthetic materials in clothing declined, alternative uses were found. One great example is the use of polyester for making beverage bottles where it replaced glass with its shatterproof properties as a significant property.

In general, it can be said that plastics enhance and even preserve life—for instance, Kevlar, when used in making canoes for recreation or when used to make a bulletproof vest. Polyester enhances life, when this highly nonreactive material is used to make replacement human blood vessels or even replacement skin for burn victims. With all the benefits attributed to plastics, they have their negative side. A genuine environmental problem exists due to the fact that the synthetic polymers do not break down easily compared with the natural polymers. Hence, there is a need not only to develop biodegradable plastics but also to work on more effective means of recycling. A lot of research is needed to study the methods of degradation and stabilization of polymers in order to design polymers according to the end-use.

Among the most important and versatile of the hundreds of commercial plastics is polyethylene. Polyethylene is used in a wide variety of applications because it can be produced in various forms. The first type to be commercially exploited was called low-density polyethylene (LDPE). This polymer is characterized by a large degree of branching, forcing the molecules to pack together rather than loosely forming a low-density material. LDPE is soft and pliable and has applications ranging from plastic bags, containers, textiles, and electrical insulation, to coatings for packaging materials.

Another form of polyethylene differing from LDPE in structure is high density polyethylene (HDPE). HDPE demonstrates little or no branching, resulting in the molecules to be tightly packed. HDPE is much more rigid than LDPE and is used in applications where rigidity is important. Major uses of HDPE are plastic tubing, bottles, and bottle caps. Other variations of polyethylene include high and ultrahigh molecular mass ones. These types are used in applications where extremely tough and resilient materials are needed.

Rubber is the most important of all elastomers. Natural rubber is obtained from the bark of the rubber tree and has been used by humans for many centuries. It is a polymer with repeating units of isoprene. In 1823, rubber was vulcanized with sulfur whilst heated and this process made it to become the valuable material it is today. In this process, sulfur chain fragments attack the rubber polymer chains, which leads to cross-linking. Most of the rubber used in the world is a synthetic variety called styrene-butadiene rubber.

Natural polymers unlike the synthetic ones do possess very complex structure. Natural polymers such as cellulose, wool, and natural rubber are used in many products in large proportions. Cellulose derivatives are one of the most versatile groups of regenerated materials with various fields of application. Cellulose is found in nature in all forms of plant life, particularly in wood and cotton. The purest form of cellulose is obtained from the seed hairs of the cotton plant that contains up to 95% cellulose. The first cellulose derivatives came to the stage around 1845 when the nitration of starch and paper led to discovery of cellulose nitrate. In 1865, for the first time, a moldable thermoplastic made of cellulose nitrate and castor oil.

In 1865, the first acetylation of cellulose was carried out, but the first acetylation process for use in industry was announced in 1894. In 1905, an acetylation process was introduced which yielded a cellulose acetate soluble in the cheap solvent acetone. It was during World War I when cellulose acetate dope found importance for weather proofing and stiffening the fabric of aircraft wings. There was a large surplus production capacity after the war, which led to civilian end uses such as the production of cellulose acetate fibers by 1920s. Cellulose acetate became the main thermoplastic molding material when the first modern injection molding machines were designed. Among the cellulose derivatives, cellulose acetates are produced in the largest volume. Cellulose acetate can be made into fibers, transparent films, and the less substituted derivatives are true thermoplastics. Cellulose acetates are moldable and can be fabricated by the conventional processes. They have toughness, good appearance, capable of many color variations including white transparency.

New applications are being developed for polymers at a very fast rate across the world at various research centers. Examples of these include electroactive polymers, nanoproducts, robotics, etc. Electroactive polymers are special types of materials that can be used for example as artificial muscles and facial parts of robots or even in nanorobots. These polymers change their shape when activated by electricity or even by chemicals. They are light weight but can bear a large force which is very useful when being utilized for artificial muscles. Electroactive polymers together with nanotubes can produce very strong actuators. Currently, research works are carried out to combine various types of electro active polymers with carbon nanotubes to make the optimal actuator. Carbon nanotubes are very strong, elastic, and conduct electricity. When they are used as an actuator, in combination with an electroactive polymer, the contractions of the artificial muscle can be controlled by electricity. Already works are under way to use electroactive polymers in space. Various space agencies are investigating the possibility of using these polymers in space. This technology has a lot to offer for the future, and with the ever increasing work on nanotechnology, electro active materials will play very important part in modern life.

This new volume presents leading-edge research in the rapidly changing and evolving field of polymer science as well as on chemical processing. The topics in the book reflect the diversity of research advances in the production and application of modern polymeric materials and related areas, focusing on the preparation, characterization, and applications of polymers. This volume also covers various manufacturing techniques. The book will help fill the gap between theory and practice in industry.

This new book

- is a collection of articles that highlights some important areas of current interest in polymer products and chemical processes.
- gives an up-to-date and thorough exposition of the present state of the art of polymer analysis.
- familiarizes the reader with new aspects of the techniques used in the examination of polymers, including chemical, physico-chemical and purely physical methods of examination.
- describes the types of techniques now available to the polymer chemist and technician, and discusses their capabilities, limitations, and applications.
- provides a balance between materials science and mechanics aspects, basic and applied research, and high-technology, and high-volume (low-cost) composite development.

The book introduces current state-of-the-art technology in modern materials with an emphasis on the rapidly growing technologies. It takes a unique approach by presenting specific materials and then progresses into a discussion of the ways in which these materials and processes are integrated into today's functioning manufacturing industry. It follows a more quantitative and design-oriented approach than other texts in the market, helping readers gain a better understanding of important concepts. Readers will also discover how material properties relate to the process variables in a given process as well as how to perform quantitative engineering analysis of manufacturing processes.

FOREWORD

It is a pleasure to write the foreword to this "special AAP Research Notes" on *Applied Research on Polymer Composites* dedicated to Professor Gennady E. Zaikov. Gennady's name is, of course, familiar to every scientist and practising chemist from his pioneering works. From a personal perspective, Gennady has had a major impact on all of our lives, taking us under his wing in one role or another during the formative stages of our academic careers. Even though he formally retired many years ago, he maintains a tireless enthusiasm for chemistry, writing books and original papers, reviewing articles, attending chemistry meetings, and giving invited lectures around the world. I am delighted to pay tribute to Gennady for his friendship, his role as a teacher and mentor, and his important and extensive contributions to polymer chemistry. This "AAP Research Note" is a token of our appreciation of a truly outstanding scientist, Professor Gennady E. Zaikov.

Gennady Efremovich Zaikov was born in Omsk, Siberia, where he also graduated from their primary, middle, and high school. He also graduated from a musical school where he studied violin. However, his parents Efrem and Matrena, decided that it might be better for their son to continue his education by following in the footsteps of his mother—a chemistry teacher in high school and Omsk's medical institute. His father was a mathematician and land-surveyor. Therefore, in 1952, Gennady moved to Moscow where he entered the Moscow State University (MSU), and he graduated with a chemistry degree in December 1957. His bachelor's degree dealt with the problem of separating Li6 and Li7 isotopes. After this, he joined the Institute of Chemical Physics (ICP) in Moscow in February 1958.

Gennady was originally invited to ICP by Professor Nikolai Markovich Emanuel. Under his guidance, G. E. Zaikov defended his PhD thesis titled "Comparison of the Kinetics and Mechanism of Oxidation of the Organic Compounds in Gaseous and Liquid Phases" in 1963. In 1968, he defended a Doctor of Science thesis titled "The Role of Media in Radical-Chain Oxidation Reactions". In 1970, he became a full professor. In 1966, Gennady became involved with polymer science. N. M. Emanuel asked

Zaikov to work on problems associated with aging and stabilization of polymers, and, later, with the combustion of polymeric materials. At that time in the 1970s, there were about 1000 scientists in the USSR working on these problems including 200 scientists from ICP under Zaikov's leadership. The research was conducted on all aspects of these polymer problems: thermal degradation, oxidation, ozonolysis, photodegradation and radiation degradation, hydrolysis, biodegradation, mechanical degradation, pyrolysis, and flammability.

After "perestroika and degradation" of the USSR in 1991, the new Russian government reduced drastically the financial support of science. Therefore, G. E. Zaikov has now with him in N. M. Emanuel Institute only 20 coworkers (instead of 200 in 1970–1980s). Fortunately, he has good scientific cooperation with 28 research centers of the former USSR (Russia, Ukraine, Georgia, Belorussia, Armenia, Tadjikistan, etc.) and with some research centers of Europe (Portugal, Spain, Belgium, Czech Republic, Poland, Bulgaria, and Romania). G. E. Zaikov is an outstanding scientist with expertise in wide areas of chemistry: chemical and biological kinetics, chemistry and physics of polymers, history of chemistry, and biochemistry.

In addition to his position at the N. M. Emanuel Institute, he is a Lecturer at the Moscow Institute of Fine Chemical Technology. He taught his students from his own books: *Degradation and Stabilization of Polymers Physical Methods in Chemistry* and *Acid Rains and Environmental Problems* G. E. Zaikov has written about 1800 original articles, 130 monographs (20 in Russian and 110 in English), and 250 chapters in 100 volumes. It is apparent from this work that he has made valuable contributions to the theory and practice of polymers—aging and development of new stabilizers for polymers, organization of their industrial production, life-time predictions for use and storage, and the mechanisms of oxidation, ozonolysis, hydrolysis, biodegradation, and decreasing of polymer flammability. New methods of polymer modification using the process of degradation were introduced into practice by Zaikov. These methods allow the production of new polymeric materials with improved properties. For a last period, he is very active in the field of semiconductors and electroconductive polymers, polymer blends, and polymer composites including nanocomposites.

G. E. Zaikov is a member of many editorial boards of journals published in Russia, Poland, Bulgaria, the United States, and England. In the former Soviet Union (after academician N. M. Emanuel's death), he head-ed the team dealing with the problem of polymer aging in the USSR and the Eastern European countries in the cooperation with the Soviet Acad-emy of Sciences. His present position is Head of Laboratory, Member of Directorium and Deputy of Department of the N. M. Emanuel Institute of Biochemical Physics Russian Academy of Sciences, and Professor of Polymer Chemistry at the Moscow State Academy of Fine Chemical Technology. His fields of interest are chemical physics, chemical kinetics, flammability, degradation and stabilization of polymers, diffusion, poly-mer materials, kinetics in biology, and history of chemistry.

On his eightieth birthday, G. E. Zaikov is in the prime of his life. Al-though support for scientists and research is now at a low point for many in Russia, he is hopeful that for the sake of his country and its future that this will improve. The practice of good science still exists in Russia, and G. E. Zaikov has been and is a significant contributor. Many of Professor Zaikov's students are currently in leadership positions around the nation and around the globe. Some of them are authors of the papers included in this issue covering a wide range of topics related to polymer composites. The topics are very broad ranging and cover industrial and practical issues. This broad range of topics as well as the outstanding nature of the research and publications is a direct measure of the impact that Professor Zaikov has had and continues to have on research and development in the field of polymer. I am truly delighted to dedicate this volume to Gennady on behalf of the chemists and polymer science community as a thank you for all of his current and future contributions to the community.

Professor A. K. Haghi, PhD
*Member of Canadian Research and Development Center of
Sciences and Cultures, Montreal, Canada*

This peer-reviewed volume is dedicated to Professor G. E. Zaikov in celebration of his eightieth birthday and his distinguished career achieve-ments. Professor Zaikov is a preeminent scientist of our times with numer-ous groundbreaking and seminal works. He is a pioneer, leading researcher and instigator of several important research directions that include theory and practice of polymers—aging and development of new stabilizers for polymers, organization of their industrial production, life-time predictions for use and storage, and the mechanisms of oxidation, ozonolysis, hydrolysis, biodegradation, and decreasing of polymer flammability., among many others. Professor Zaikov's research accomplishments are incredibly broad in both subjects and research methodologies. The topics of his research span a wide spectrum of scientific areas including nanocomposites, polymers, elasticity and structures, and material science, just to name a few. His research aspects involve modeling, hard and soft analysis, the design of ingenious, efficient and effective computational schemes, rigorous numerical analysis and error estimations, or whatever the problem at hand calls for. Professor Zaikov's publications speak vividly of his distinguished research career.

Ali Pourhashemi, PhD
*Professor, Department of Chemical and Biochemical Engineering,
Christian Brothers University, Memphis, Tennessee, USA*

It is our great privilege and pleasure to introduce this volume, which is dedicated to Professor Gennady E. Zaikov on the occasion of his eightieth birthday. Professor Gennady E. Zaikov has made many outstanding contributions to the field of polymer science, particularly for the conservation of biodiversity. His research has been based on various aspects of modern materials, offering valuable contributions to modern chemistry, food chemistry, and more recently, nanotechnology. During his distinguished and productive career, Gennady E. Zaikov has either authored or coauthored over 200 peer-reviewed scientific books and a great number of contributions in congress proceedings. This volume, on the occasion of his eightieth birthday on January 7, 2015, represents a tribute to his outstanding scientific contributions and an opportunity to express congratulations and best wishes from all of us, colleagues, and friends. Our personal thanks go also to all the authors and reviewers who have contributed to the success of this special issue.

Abbas Hamrang, PhD
*Professor and Senior Polymer Scientist,
Independent Polymer Consultant, Manchester, UK*

This volume is dedicated to Professor G. E. Zaikov, Emeritus Professor of Russian Academy of Sciences, on the occasion of his eightieth birthday. He is a distinguished Russian scientist and one of the most eminent scientists of the last half-century. He has contributed much to the field of semiconductors and electroconductive polymers, polymer blends, and polymer composites including nanocomposites. He continues to contribute to this day. He is also an active member of many editorial board for international journals. With the publication of this volume, we wish to pay homage to his great scientific achievements and to thank him for his friendship and kindness. We are honored to recognize Professor Zaikov's great contribution to the scientific world and wish him more fruitful experiences in his further research. We also warmly thank all the authors for their kind cooperation on the production of this volume.

Pooria Pasbakhsh, PhD
School of Engineering, Monash University
Sunway Campus, Malaysia

Professor Zaikov has been an enthusiastic promoter and practitioner of global scientists, training and collaborating with students and researchers of different nationalities and of diverse cultural or educational backgrounds. He has constantly and unselfishly contributed to the global dissemination of research in polymer materials, and his outstanding research leadership and tireless service have positively impacted the worldwide prospering of research in these areas.

The contributing authors of this special volume consist mostly of Professor Zaikov's former PhD students, postdoctoral fellows, and colleagues who at one time or another had the opportunity to collaborate with Professor Zaikov on various research projects. This collection of research papers is a testimony of Professor Zaikov's wide scope of research accomplishment and profound influence in the international research community on polymer composites and on scientific modeling. No doubt that great thoughts and ideas will continue to stream out of Professor Zaikov's beautiful mind, bringing his already distinguished career to an even higher plateau.

Sabu Thomas, PhD
Director, School of Chemical Sciences, Professor of Polymer Science &
Technology & Honorary Director of the International and Inter
University Centre for Nanoscience and Nanotechnology,
Mahatma Gandhi University, Kottayam, India

CHAPTER 1

TRENDS IN NEW GENERATION OF BIODEGRADABLE POLYMERS (PART 1)

A.L. IORDANSKII, G.A. BONARTSEVA, YU.N. PANKOVA, S.Z. ROGOVINA, K.Z. GUMARGALIEVA, G.E. ZAIKOV, and A.A. BERLIN

CONTENTS

1.1   INTRODUCTION

Currently, the intensive development of biodegradable and biocompatible materials for medical implication provokes comprehensive interdisciplinary studies on biopolymer structures and functions. The well-known and applicable biodegradable polymers are polylactides (PLAs) and polyglycolides (PGA), and their copolymers, poly-ε-caprolactone, poly(orthoesters), poly-β-malic acid, poly(propylene fumarate), polyalkylcyanoacrylates, polyorthoanhydrides, polyphosphazenes, poly(propylene fumarate), some natural polysaccharides (starch, chitosan, alginates, agarose, dextran, chondroitin sulfate, and hyaluronic acid), and proteins (collagen, silk fibroin, spidroin, fibrin, gelatin, and albumin). As some of these polymers should be synthesized through chemical stages (e.g. via lactic and glycolic acids), it is not quite correct to define them as the biopolymers.

Besides biomedicine applications, the biodegradable biopolymers attract much attention as perspective materials in wide areas of industry, nanotechnologies, farming, and packaging owing to the relevant combination of biomedical, transport, and physical-chemical properties. It is worth to emphasize that only medical area of these biopolymers includes implants and prosthesis, tissue engineering scaffolds, novel drug dosage forms in pharmaceutics, novel materials for dentistry, and others.

Each potentially applicable biopolymer arranges a wide multidisciplinary network, which usually includes tasks of searching for efficacy ways of biosynthesis reactions; economical problems associated with large-scale production; academic studies of mechanical, physicochemical, biochemical properties of the polymer and material of interest; technology of preparation and using this biopolymer; preclinical and clinical trials of these materials and products; a market analysis and perspectives of application of the developed products and many other problems.

Poly(R)-3-hydroxybutyrate) (PHB) is an illustrative example for one of centers for the formation of the above-mentioned scientific-technological network and a basis for the development of various biopolymer systems [1, 2]. In recent decades, an intense development of biomedical application of bacterial PHB in the production of biodegradable polymer implants and controlled drug release systems [3–6] needs for comprehensive understanding of the PHB biodegradation process. Examination of PHB degradation process is also necessary for development of novel friendly environment polymer packaging [7–9]. It is generally accepted that biodegradation of PHB both in living systems and in environment occurs via enzymatic and nonenzymatic processes that take place simultaneously under natural conditions. It is, therefore, important to understand both processes [6, 10]. Opposite to other biodegradable polymers (e.g. PGA and PLGA), PHB is considered to be moderately resistant to degradation *in vitro* as well as to biodegradation in biological media. The rates of degradation are influenced by the characteristics of the polymer, such as chemical composition, crystallinity, morphology and molecular weight [11, 12]. Despite that PHB application *in vitro* and *in vivo* has been intensively investigated, the most of the available data are often incomplete and sometimes even contradictory. The presence of conflicting data can be partially explained by the fact that biotechnologically produced PHB with standardized properties is relatively rare and is not readily available due to a wide variety of its biosynthesis sources and different manufacturing processes.

Above-mentioned inconsistencies can be explained also by excess applied trend in PHB degradation research. At most of the papers observed in this review, PHB degradation process has been investigated in the narrow framework of development of specific medical devices. Depending on the applied biomedical purposes, biodegradation of PHB was investigated under different geometries: films and plates with various thickness [13–16], cylinders [17–19], monofilament threads [20–22] and micro- and nanospheres [23, 24]. In these experiments, PHB was used from various sources, with different molecular weights and crystallinities. Besides, different technologies of PHB manufacture of devices affect important characteristics such as polymer porosity and surface structure ([14, 15]. Reports related to the complex theoretical research of mechanisms of hydrolysis, enzymatic degradation, and biodegradation *in vivo* of PHB processes are relatively rare [13–15, 16, 25–27], which attaches great value and importance to these investigations. Nevertheless, the effect of thickness, size and geometry of PHB device, molecular weight, and crystallinity of PHB on the mechanism of PHB hydrolysis and biodegradation was not yet well clarified.

1.2   HYDROLYSIS AND ENZYMATIC DEGRADATION OF PHB

1.2.1   NONENZYMATIC HYDROLYSIS OF PHB IN VITRO

Examination of hydrolytic degradation of natural poly((R)-3-hydroxybu-tyrate) *in vitro* is a very important step for understanding of PHB bio-degradation. There are several very profound and careful examinations of PHB hydrolysis that was carried out for 10–15 years [25–28]. Hydrolytic degradation of PHB was usually examined under standard experimental conditions simulating internal body fluid: in buffered solutions with pH = 7.4 at 37°C but at seldom cases, the higher temperature (55°C, 70°C, and more) and other values of pH (from 2 to 11) were selected.

The classical experiment for examination of PHB hydrolysis in com-parison with hydrolysis of another widespread biopolymer, polylactic acid (PLA), was carried out by Koyama N. and Doi Y. [25]. They compared films (10 × 10 mm^2 size, 50 μm thickness, 5 mg initial mass) from PHB (M_n = 3,00,000, M_w = 6,50,000) with polylactic acid (M_n = 9,000, M_w = 21,000) prepared by solvent casting and aged for 3 weeks to reach equilib-rium crystallinity. They have shown that hydrolytic degradation of natural PHB is a slow process. The mass of PHB film remained unchanged at 37°C in 10 mM phosphate buffer (pH = 7.4) over a period of 150 days, while the mass of the PLA film rapidly decreased with time and reached 17 per cent of the initial mass after 140 days. The rate of decrease in the M_n of the PHB was also much slower that in the M_n of PLA.

The M_n of the PHB decreased approximately to 65 per cent of the ini-tial value after 150 days, while the M_n of the PLA decreased to 20 per cent (M_n = 2,000) at the same time. As PLA used at this research was with low molecular weight, it is worth to compare these data with the data of hy-drolysis investigation with the same initial M_n as for PHB. In other works, the mass loss of two polymer films (PLA and PHB) with the same thick-ness (40 μm) and molecular weight (M_w = 450,000) was studied *in vitro*. It was shown that the mass of PLA film decreased to 87 per cent, whereas the mass of PHB film remained unchanged at 37°C in 25 mM phosphate buffer (pH = 7.4) over a period of 84 days, but after 360 days the mass of PHB film was 64.9 per cent of initial one [29–31].

The cleavage of polyester chains is known to be catalyzed by the car-boxyl end groups, and its rate is proportional to the concentrations of wa-ter, and ester bonds those on the initial stage of hydrolysis are constant owing to the presence of a large excess of water molecules and ester bonds of polymer chains. Thus, the kinetics of nonenzymatic hydrolysis can be expressed by a simplified equation [32-33]:

$$\ln M_n = \ln M_n^0 - k_h t \qquad (1)$$

where M_n and M_n^0 are the number-average molecular weights of a polymer component at time t and 0, respectively and k_h is the effective hydrolysis constant.

The average number of bond cleavage per original polymer molecule, N, is given by equation 2:

$$N = (M_n^0/M_n) - 1 = k_d M_m P_n^0 t, \qquad (2)$$

where k_d is the effective rate constant of hydrolytic depolymerization, and P_n^0 is the initial number-average degree of polymerization at time 0, M_m is the constant molecular mass of monomer. Thus, if the chain scission is completely random, the value of N is linearly dependent on time.

The molecular weight decrease with time is the distinguishing feature of mechanism under nonenzymatic hydrolysis condition in contrast to enzymatic hydrolysis condition of PHB when M_n values remained almost unchanged. It was supposed also that water-soluble oligomers of PHB with molecular mass about 3 kDa may accelerate the hydrolysis rate of PHB homopolymer [25]. In contrast, Freier et al. [14] showed that PHB hydrolysis was not accelerated by the addition of predegraded PHB: the rate of mass and M_w loss of blends (70/30) from high-molecular PHB (M_w = 6,41,000) and low-molecular PHB (M_w = 3,000) was the same with degradation rate of pure high-molecular PHB. Meanwhile, the addition of amorphous atactic PHB (at PHB) (M_w = 10,000) to blend with high-molecular PHB caused significant acceleration of PHB hydrolysis: the relative mass loss of PHB/atPHB blends was 7 per cent in comparison with 0 per cent mass loss of pure PHB; the decrease of M_w was 88 per cent in comparison with 48 per cent M_w decrease of pure PHB [14, 34]. We have showed that the rate of hydrolysis of PHB films depends on M_w of PHB. The PHB films of high molecular weight (M_w=450.000 and 1,000,000) degraded slowly as it was described earlier, whereas films from PHB of low molecular weight (M_w = 150,000 and 300,000 kDa) lost weight relatively gradually and more rapidly [29–31].

To enhance the hydrolysis of PHB, a higher temperature was selected for degradation experiments: 55°C, 70°C, and more [25]. It was showed by the same research team that the weights of films (12 mm diameter, 65 μm thick) from PHB (M_n = 768 and 22 kDa, M_w = 1,460 and 75 kDa) were unchanged at 55°C in 10 mM phosphate buffer (pH = 7.4) over a period of 58 days. The M_n value decreased from 768 to 245 kDa for 48 days. The film thickness increased from 65 to 75 μm for 48 days, suggesting that water permeated the polymer matrix during the hydrolytic degradation. Examination of the surface and cross section of PHB films before and after hydrolysis showed that surface after 48 days of hydrolysis was apparently unchanged, while the cross section of the film exhibited a more porous structure (pore size < 0.5 μm). It was also shown that the rate of hydrolytic degradation is not dependent on the crystallinity of PHB film. The observed data indicates that the nonenzymatic hydrolysis of PHB in the aqueous media proceeds via a random bulk hydrolysis of ester bonds in the polymer chain films and occurs throughout the whole film, as water permeates the polymer matrix [25–26]. Moreover, as over the whole degradation time the first-order kinetics was observed and the molecular weight distribution was unimodal, a random chain scission mechanism is very probable both on the crystalline surfaces and in the amorphous regions of the biopolymer [14, 35]. For synthetic amorphous at PHB, it was shown that its hydrolysis is a two-step process. First, the random chain scission proceeds that accompanying with a molecular weight decrease. Then, at a molecular weight of about 10,000, mass loss begins [28].

The analysis of literature data shows a great spread in values of rate of PHB hydrolytic degradation *in vitro*. It can be explained by different thickness values of PHB films or geometry of PHB devices used for experiment as well as by different sources, purity degree, and molecular weight of PHB (Table 1.1). At 37°C and pH = 7.4, the weight loss of PHB (unknown M_w) films (500 μm thick) was 3 per cent after 40 days incubation [38], 0 per cent after 52 weeks (364 days) and after 2 years (730 days) incubation (640 kDa PHB, 100 μm films) [14–15], 0 per cent after 150 days incubation (650 kDa PHB, 50 μm film) [25], 7.5 per cent after 50 days incubation (279 kDa PHB, unknown thickness of films) [37], 0 per cent after 3 months (84 days) incubation (450 kDa PHB, 40 μm films), 12 per cent after 3 months (84 days) incubation (150 kDa PHB, 40 μm films) [29–30], 0 per cent after 180 days incubation of monofilament threads (30 μm in diameter) from PHB (470 kDa) [22–23]. The molecular weight of PHB dropped to 36 per cent of the initial values after 2 years (730 days) of storage in buffer solution [15], to 87 per cent of the initial values after 98 days [38], to 58 per cent of the initial values after 84 days [29–30] (Table 1.1).

TABLE 1.1　Nonenzymatic hydrolysis of PHB *in Vitro*

(Data for Comparison)

Type of device	Initial M_w (kDa)	Size/thickness (μm)	Condi-tions	Relative mass loss (%)	Relative loss of M_w (%)	Time, days	Links
Film	650	50	37 C, pH = 7.4	0	35	150	25
Film	640	100	37°C, pH = 7.4	0	64	730	15
Film	640	100	37°C, pH = 7.4	0	45	364	14
Film	450	40	37 C, pH = 7.4	0	42	84	29-30
Film	150	40	37°C, pH = 7.4	12	63	84	29-30
Film	450	40	37°C, pH = 7.4	35,1	-	360	31
Film	279	-	37°C, pH = 7.4	7.5	-	50	36
Plate	-	500	37°C, pH = 7.4	3	-	40	30
Plate	380	1,000	37°C, pH = 7.4	0	-	28	43
Plate	380	2,000	37°C, pH = 7.4	0	8	98	43
Thread	470	30	37°C, pH = 7.0	0	-	180	23
Thread	-	-	37°C, pH = 7.2	0	-	182	22

TABLE 1.1   *(Continued)*

Type of device	Initial M_w (kDa)	Size/thickness (µm)	Condi-tions	Relative mass loss (%)	Relative loss of M_w (%)	Time, days	Links
Micro-spheres	50	250-850	37°C, pH = 7.4	0	0	150	42
Thread	470	30	37°C, pH = 5.2	0	-	180	22
Film	279	-	37°C, pH = 10	100	-	28	36
Film	279	-	37°C, pH = 13	100	-	19	36
Film	650	50	55°C, pH = 7.4	0	68	150	25
Plate	380	2,000	55 C, pH = 7.4	0	61	98	43
Film	640	100	70°C, pH = 7.4	-	55	28	14
Film	150	40	70°C, pH = 7.4	39	96	84	29–30
Film	450	40	70°C, pH = 7.4	12	92	84	29–30
Micro-spheres	50	250-850	85°C, pH = 7.4	50	68	150	42
Micro-spheres	600	250-850	85°C, pH = 7.4	25	-	150	42

    In acidic or alkaline aqueous media, PHB degrades more rapidly: 0 per cent degradation after 140 days incubation in 0.01 M NaOH (Ph = 11) (200 kDa, 100 µm film thickness) with visible surface changing [39], 0 per cent degradation after 180 days incubation of PHB threads in phosphate buffer (pH=5.2 and 5.9) [23], and complete PHB films biodegradation after 19 days (pH = 13) and 28 days (pH=10) [37]. It was demonstrated that after 20 weeks of exposure to NaOH solution, the surfaces of PHB samples became rougher, along with an increased density in their surface layers. From these results, one may surmise that the nonenzymatic degradation of PHAs progresses on their surfaces before noticeable weight loss occurs as illustrated in Figure 1.1 [39].

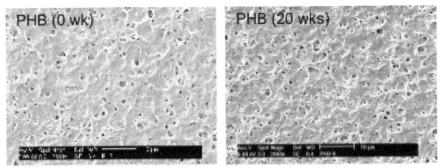

FIGURE 1.1   Scanning electron microscopic photographs of PHB films both before (0 wk, panels on the left) and after 20 weeks (panels on the right) of nonenzymatic hydrolysis in 0.01 N NaOH solution (scale bars, 10 μm) [39].

It was also shown that treatment of PHB film with 1M NaOH caused a decrease in pore size on film surface from 1–5 μm to around 1 μm, which indicates a partially surface degradation of PHB in alkaline media [40–41]. At higher temperature, no weight loss of PHB films and threads was observed after incubation of 98 and 182 days in phosphate buffer (pH=7.2) at 55°C and 70°C, respectively [22], 12 per cent and 39 per cent of PHB (450 and 150 kDa, respectively) films after 84 days incubation at 70°C [35, 40], 50 per cent and 25 per cent after 150 days incubation of microspheres (250–850 μm diameter) from PHB (50 kDa and 600 kDa, respectively) [42].

During degradation of PHB monofilament threads, films, and plates, the change of mechanical properties was observed under different conditions *in vitro* [22, 43]. It was shown that a number of mechanical indices of threads became worse: load at break lost 36 per cent, strain at break lost 33 per cent, Young's modulus didn't change, tensile strength lost 42 per cent after 182 days incubation in phosphate buffer (pH=7.2) at 70°C. However, at 37°C the changes were more complicated: at first load at break increased from 440 to 510 g (16 per cent) at 90th day and then decreased to the initial value on 182nd day, strain at break increased rapidly from 60 to 70 per cent (in 17 per cent) at 20th day and then gradually increased to

75 per cent (in 25 per cent) at 182nd day, and Young's modulus did not change [22].

For PHB films, a gradual 32 per cent decrease in Young's modulus and 77 per cent fall in tensile strength during 120-day incubation in phosphate buffer (pH = 7.4) at 37°C [43] were demonstrated. For PHB plates, more complicated changes were observed: at first, tensile strength dropped to 13 per cent at 1st day and then increased to the initial value at 28th day, Young's modulus dropped to 32 per cent for 1st day and then remain unchanged up to 28th day, stiffness also decreased sharply to 40 per cent for 1st day and then remain unchanged up to 28th day [44].

1.2.2 ENZYMATIC DEGRADATION OF PHB *IN VITRO*

The examination of enzymatic degradation of PHB *in vitro* is the following important step for understanding of PHB performance in animal tissues and environment. The most papers observed degradation of PHB by depolymerases of its own bacterial producers. The degradation of PHB *in vitro* by depolymerase was thoroughly examined and mechanism of enzymatic PHB degradation was perfectly clarified by Doi [25–26]. At these early works, it was shown that 68–85 per cent and 58 per cent mass loss of PHB (M_w=650–768 and 22 kDa, respectively) films (50–65 μm thick) occurred for 20 h under incubation at 37 °C in phosphate solution (pH = 7.4) with depolymerase (1.5–3 μg/ml) isolated from *Alcaligenes faecalis*. The rate (k_e) of enzymatic degradation of films from PHB (M_n=768 and 22 kDa) was 0.17 and 0.15 mg/h, respectively.

The thickness of polymer films dropped from 65 to 22 μm (32% of initial thickness) during incubation. The scanning electron microscopic examination showed that the surface of the PHB film after enzymatic degradation was apparently blemished by the action of PHB depolymerase, while no change was observed inside the film. Moreover, the molecular weight of PHB remained almost unchanged after enzymatic hydrolysis: the M_n of PHB decreased from 768 to 669 kDa or unchanged (22 kDa) [25–26].

The extensive literature data on enzymatic degradation of PHB by specific PHB depolymerases was collected in detail in review of Sudesh, Abe, and Doi [45]. We would like to summarize some the most important data. However, at first it is necessary to note that PHB depolymerase is very specific enzyme and the hydrolysis of polymer by depolymerase is a unique process. But in animal tissues and even in environment, the enzymatic degradation of PHB occurred mainly by nonspecific esterases [24, 46]. Thus, in the frameworks of this review, it is necessary to observe the fundamental mechanisms of PHB enzymatic degradation.

The rate of enzymatic erosion of PHB by depolymerase is strongly dependent on the concentration of the enzyme. The enzymatic degradation of solid PHB polymer is heterogeneous reaction involving two steps, namely, adsorption and hydrolysis. The first step is adsorption of the enzyme onto the surface of the PHB material by the binding domain of the PHB depolymerase, and the second step is hydrolysis of polyester chains by the active site of the enzyme. The rate of enzymatic erosion for chemosynthetic PHB samples containing both monomeric units of (R)- and (S)-3-hydrohybutyrate is strongly dependent on both the stereocomposition and tacticity of the sample as well as on substrate specificity of PHB depolymerase.

The water-soluble products of random hydrolysis of PHB by enzyme showed a mixture of monomers and oligomers of (R)-3-hydrohybutyrate. The rate of enzymatic hydrolysis for melt-crystallized PHB films by PHB depolymerase decreased with an increase in the crystallinity of the PHB film, while the rate of enzymatic degradation for PHB chains in an amorphous state was approximately 20 times higher than the rate for PHB chains in a crystalline state. It was suggested that the PHB depolymerase predominantly hydrolyzes polymer chains in the amorphous phase and then, subsequently, erodes the crystalline phase. The surface of the PHB film after enzymatic degradation was apparently blemished by the action of PHB depolymerase, while no change was observed inside the film.

Thus, depolymerase hydrolyses of the polyester chains in the surface layer of the film and polymer erosion proceed in surface layers, while dissolution and the enzymatic degradation of PHB are affected by many factors such as monomer composition, molecular weight, and degree of crystallinity [45]. The PHB polymer matrix ultrastructure [47] plays also very important role in enzymatic polymer degradation [48].

In the next step, it is necessary to observe enzymatic degradation of PHB under the conditions that modeled the animal tissues and body fluids containing nonspecific esterases. *In vitro* degradation of PHB films in the presence of various lipases as nonspecific esterases was carried out in buffered solutions containing lipases [18, 49–50], in digestive juices (for example, pancreatin) [14], simulated body fluid [51] biological media (serum, blood, etc.) [23], and crude tissue extracts containing a mixture of enzymes [24] to examine the mechanism of nonspecific enzymatic degradation process. It was noted that a Ser..His..Asp triad constitutes the active center of the catalytic domain of both PHB depolymerase [52] and lipases [53]. The serine is a part of the pentapeptide Gly X1-Ser-X2-Gly, which has been located in all known PHB depolymerases as well as in lipases, esterases, and serine proteases [52].

On the one hand, it was shown that PHB was not degraded for 100 days with a quantity of lipases isolated from different bacteria and fungi [49–50]. On the other hand, the progressive PHB degradation by lipases was shown [18, 40–41]. PHB enzymatic biodegradation was also studied in biological media: it was shown that with pancreatin addition no additional mass loss of PHB was observed in comparison with simple hydrolysis [14], and the PHB degradation process in serum and blood was demonstrated to be similar to hydrolysis process in buffered solution [31], whereas progressive mass loss of PHB sutures was observed in serum and blood: 16 per cent and 25 per cent, respectively, after 180 days incubation [23], crude extracts from liver, muscle, kidney, heart, and brain showed the activity to degrade the PHB: from 2 per cent to 18 per cent mass loss of PHB microspheres after 17 h incubation at pH 7.5 and 9.5 [24]. The weight loss of PHB (M_w = 2,85,000) films after 45 days incubation simulated body fluid was about 5 per cent [51]. The degradation rate in solution with pancreatin addition, obtained from the decrease in M_w of pure PHB, was accelerated about threefold: 34 per cent decrease in M_w after incubation for 84 days in pancreatin (10 mg/ml in Sorensen buffer) versus 11 per cent decrease in M_w after incubation in phosphate buffer [14]. The same data was obtained for PHB biodegradation in buffered solutions with porcine lipase addition: 72 per cent decrease in M_w of PHB (M_w = 4,50,000) after incubation for 84 days with lipase (20 U/mg, 10 mg/ml in Tris-buffer) versus 39 per cent decrease in M_w after incubation in phosphate buffer [18]. This observation is in contrast with enzymatic degradation by PHB depolymerases which was reported to proceed on the surface of the polymer film with an almost unchanged molecular weight [24–25]. It has been proposed that for depolymerases the relative size of the enzyme compared with the void space in solvent cast films is the limiting factor for diffusion into the polymer matrix [54], whereas lipases can penetrate into the polymer matrix through pores in PHB film [40–41]. It was shown that lipase (0.1 g/l in buffer) treatment for 24 h caused significant morphological change in PHB film surface: transferring from native PHB film with many pores ranging from 1 to 5 μm in size into a pore free surface without producing a quantity of hydroxyl groups on the film surface. It was supposed that the pores had a fairly large surface exposed to lipase, thus it was degraded more easily (Figure 1.2) [40–41]. It also indicates that lipase can partially penetrate into pores of PHB film but the enzymatic degradation proceeds mainly on the surface of the coarse polymer film achievable for lipase. Two additional effects reported for depolymerases could be of importance. It was concluded that segmental mobility in amorphous phase and polymer hydrophobicity play an important role in enzymatic PHB degradation by nonspecific esterases [14]. Significant impairment of the tensile strength and other mechanical properties were observed during enzymatic biodegradation of PHB threads in serum and blood. It was shown that load at break lost 29 per cent, Young's modulus lost 20 per cent, and tensile strength did not change after 180 days of incubation of threads, and the mechanical properties changed gradually [23].

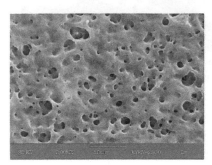

FIGURE 1.2 Scanning electron microscopy photographs documented the surface structure of PHB polymer films: (a) PHB film; (b) PHB film treated with lipase (0.1 g/l at 30°C and pH = 7.0 for 24 h) [41].

1.3 BIODEGRADATION OF PHB BY SOIL MICROORGANISMS

Polymers exposed to the environment are degraded by their hydrolysis, mechanical, thermal, oxidative, and photochemical destruction, and biodegradation [7, 38, 55, 56]. One of the valuable properties of PHB is its biodegradability, which can be evaluated using various field and laboratory tests. Requirements for the biodegradability of PHB may vary in accordance with its applications. The most attractive property of PHB with respect to ecology is that it can be completely degraded by microorganisms finally to CO_2 and H_2O. This property of PHB allows to manufacture biodegradable polymer objects for various applications (Figure 1.3) [4].

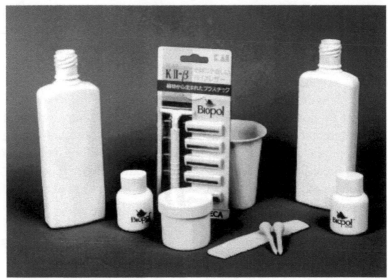

FIGURE 1.3   Molded PHB objects for various applications. In soil burial or composting experiments, such objects biodegrade in about 3 months [4].

The degradation of PHB and its composites in natural ecosystems, such as soil, compost, and bodies of water, was described in a number of publications [4, 38, 55, 56]. Maergaert et al. isolated from soil more than 300 microbial strains capable of degrading PHB *in vitro* [55]. The bacteria detected on the degraded PHB films were dominated by the genera *Pseudomonas*, *Bacillus*, *Azospirillum, Mycobacterium*, and *Streptomyces*, and so on. The samples of PHB have been tested for fungicidity and resistance to fungi by estimating the growth rate of test fungi from the genera *Aspergillus*, *Aureobasidium*, *Chaetomium*, *Paecilomyces*, *Penicillum*, *Trichoderma* under optimal growth conditions. PHB film did exhibit neither fungicide properties nor the resistance to fungal damage, and served as a good substrate for fungal growth [57].

Biodegradability of PHB films under aerobic, microaerobic, and anaerobic conditions in the presence and absence of nitrate by microbial populations of soil, sludge from anaerobic and nitrifying/denitrifying reactors, and sediment of a sludge deposit site, as well as to obtain active denitrifying enrichment culture degrading PHB (Figure 1.4) [58] were studied. Changes in molecular mass, crystallinity, and mechanical properties of PHB have been studied. A correlation between the PHB degradation degree and the molecular weight of degraded PHB was demonstrated.

The most degraded PHB exhibited the highest values of the crystallinity index. As it has been shown by Spyros et al., PHAs contain amorphous and crystalline regions, of which the former are much more susceptible to microbial attack [59]. If so, the microbial degradation of PHB must be associated with a decrease in its molecular weight and an increase in its crystallinity, which was really observed in the experiments. Moreover, microbial degradation of the amorphous regions of PHB films made them more rigid. However, further degradation of the amorphous regions made the structure of the polymer much looser [58].

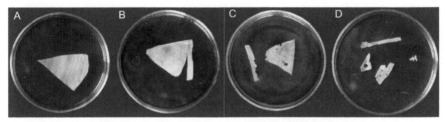

FIGURE 1.4 Undegraded PHB film (A) and PHB films with different degrees of degradation after 2 months incubation in soil suspension: anaerobic conditions without nitrate (B), microaerobic conditions without nitrate (C), and microaerobic conditions with nitrate (D) [58].

PHB biodegradation in the enriched culture obtained from soil on the medium used to cultivate denitrifying bacteria (Gil'tai medium) has also been studied. The dominant bacterial species, *Pseudomonas fluorescens* and *Pseudomonas stutzeri*, have been identified in this enrichment culture. Under denitrifying conditions, PHB films were completely degraded for 7 days. Both the film weight and M_w of PHB decreased with time. In contrast to the data of Doi et al. [26] who found that M_w of PHB remained unchanged upon enzymatic biodegradation in an aquatic solution of PHB depolymerase from *Alcaligenes faecalis*, in our experiments, the average viscosity molecular weight of the higher and lower molecular polymers decreased gradually from 1,540 to 580 kDa and from 890 to 612 kDa, respectively. As it was shown at single PHB crystals [47], the "exo"-type cleavage of the polymer chain, that is, a successive removal of the terminal groups, is known to occur at a higher rate than the "endo"-type cleavage, that is, a random breakage of the polymer chain at the enzyme-binding sites. Thus, the former type of polymer degradation is primarily responsible for changes in its average molecular weight. However, the "endo"-type attack plays an important role during the initiation of biodegradation, because at the beginning, a few polymer chains are oriented so that their ends are accessible to the effect of the enzyme [60]. Biodegradation of the lower molecular polymer, which contains a higher number of terminal groups, is more active, probably, because the "exo"-type degradation is more active in lower than in higher molecular polymer [58, 61].

1.4   BIODEGRADATION OF PHB *IN VIVO* IN ANIMAL TISSUES

The first scientific works on biodegradation of PHB *in vivo* in animal tissues were carried out 15–20 years ago by Miller et al., and Saito et al [22, 24]. They are high-qualitative research studies that disclosed many important characteristics of this process. As it was noted earlier, both enzymatic and nonenzymatic processes of biodegradation of PHB *in vivo* can occur simultaneously under normal conditions. However, it does not mean that polymer biodegradation *in vivo* is a simple combination of nonenzymatic hydrolysis and enzymatic degradation. Moreover, *in vivo* biodegradation (decrease of molecular weight and mass loss) of PHB) is a controversial subject in the literature. As it was noted above for *in vitro* PHB hydrolysis, the main reason for the controversy is the use of samples made by various processing technologies and the incomparability of different implantation and animal models. Most of the research studies on PHB biodegradation were carried out with the use of prototypes of various medical devices on the base of PHB: solid films and plates [13, 16, 18, 31, 62], porous patches [14, 15], porous scaffolds [63], electrospun microfiber mats [64], nonwoven patches consisted of fibers [65–69], screws [31], cylinders as nerve guidance channels and conduits [16, 20, 21], monofilament sutures [22–23], cardiovascular stents [70], microspheres [24,71]. *In vivo* biodegradation was studied on various laboratory animals: rats [14, 18, 20–24, 64], mice [16, 75], rabbits [13, 62, 70, 72], minipigs [15], cats [20], calves [65], sheep [66–68], and even at clinical trials on patients [69]. It is obviously that these animals differ in the level of metabolism very much: for example, only weight of these animals differs from 10–20 g (mice) to 50 kg (calves). The implantation of devices from PHB was carried out in different ways: subcutaneously [13, 16, 18, 22, 23, 72], intraperitoneally on a bowel [14], subperiostally on the osseus skull [15–62], nerve wrap-around [19–21], intramuscularly [71–72], into the pericardium [66–69], into the atrium [65], and intravenously [24]. The terms of implantation were also different: 2.5 h, 24 h, 13 days, 2 months [24]; 7, 14, 30 days [21], 2, 7, 14, 21, 28, 55, 90, 182 days [22]; 1, 3, 6 months [13, 16, 19]; 3, 6, 12 months [65]; 6, 12 months [66]; 6, 24 months [69]; 3, 6, 9, 12, 18, 24 months [68].

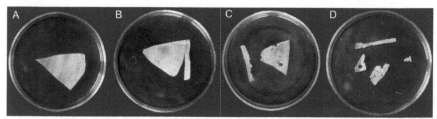

FIGURE 1.5 Undegraded PHB film (A) and PHB films with different degrees of degradation after 2 months incubation in soil suspension: anaerobic conditions without nitrate (B), microaerobic conditions without nitrate (C), and microaerobic conditions with nitrate (D) [58].

The entire study of PHB *in vivo* biodegradation was done by Go-golewski et al. and Qu et al. [13, 16]. It was shown that PHB lost about 1.6 per cent (injection-molded film, 1.2 mm thick, M_w of PHB=130 kDa) [16] and 6 per cent (solvent-casting film, 40 μm thick, M_w=534 kDa) [13] of initial weight after 6 months of implantation. However, the observed small weight loss was partially due to the leaching out of low molecular weight fractions and impurities present initially in the implants. The M_w of PHB decreased from 130,000 to 74,000 (57% of initial M_w) [16] and from 534,000 to 216,000 (40% of initial Mw) [13] after 6 months of implanta-tion. The polydispersity of PHB polymers narrowed during implantation. PHB showed a constant increase in crystallinity (from 60.7 to 64.8%) up to 6 months [16] or an increase (from 65.0 to 67.9%) after 1 month and a fall again (to 64.5%) after 6 months of implantation [13], which sug-gests that the degradation process had not affected the crystalline regions. These data are in accordance with those of PHB hydrolysis [25] and enzymatic PHB degradation by lipases *in vitro* [14], where M_w decrease was observed. The initial biodegradation of amorphous regions of PHB *in vivo* is similar to PHB degradation by depolymerase [45].

Thus, the observed biodegradation of PHB showed coexistence of two different degradation mechanisms in hydrolysis in the polymer: enzymatically or nonenzymatically catalyzed degradation. Although nonenzymatical catalysis occurred randomly in homopolymer, indicated by M_w loss rate in PHB, at some point in a time, a critical molecular weight is reached whereupon enzyme-catalyzed hydrolysis accelerated degradation at the surface because easier enzyme/polymer interaction becomes possible.

However, considering the low weight loss of PHB, the critical molecular weight appropriate for enzymes predominantly does not reach, yet resulting low molecular weight and crystallinity in PHB could provide some sites for the hydrolysis of enzymes to accelerate the degradation of PHB [13, 16]. Additional data revealing the mechanism of PHB biodegradation in animal tissues was obtained by Kramp et al. in long-term implantation experiments. A very slow, clinically not recordable degradation of films and plates was observed during 20 months (much more than in experiments mentioned earlier).

A drop in the PHB weight loss evidently took place between the 20th and 25th months. Only initial signs of degradation were to be found on the surface of the implant until 20 months after implantation, but no more test body could be detected after 25 months [62]. The complete biodegradation *in vivo* in the wide range from 3 to 30 months of PHB was shown by other research studies [65, 67–69, 73], whereas almost no weight loss and surface changes of PHB during 6 months of biodegradation *in vivo* was shown [16, 22]. Residual fragments of PHB implants were found after 30 months of the implantation of patches [66, 68]. A reduction of PHB patch size in 27 per cent was shown in patients after 24 months after surgical procedure on pericardial closure with the patch [69].

Significantly more rapid biodegradation *in vivo* was shown by other research studies [13, 20, 23, 43, 65]. It was shown that 30 per cent mass loss of PHB sutures occurred gradually during 180 days of *in vivo* biodegradation with minor changes in the microstructure on the surface and in volume of sutures [23]. It was shown that PHB nonwoven patches (made to close atrial septal defect in calves) was slowly degraded by polynucleated macrophages, and 12 months postoperatively no PHB device was identifiable but only small particles of polymer were still seen. The absorption time of PHB patches was long enough to permit regeneration of a normal tissue [65]. The progressive biodegradation of PHB sheets was demonstrated qualitatively at 2, 6, and 12 months after implantation as weakening of the implant surface, tearing/cracking of the implant, fragmentation, and a decrease in the volume of polymer material [21, 43, 72].

The complete biodegradation of PHB (M_w = 150 − 1,000 kDa) thin films (10–50 µm) for 3–6 months was shown and degradation process was described. The process of PHB biodegradation consists of several phases. At initial phase, PHB films was covered by fibrous capsule. At second phase, capsulated PHB films very slowly lost weight with simultaneous increase of crystallinity and decrease of M_w and mechanical properties of PHB. At third phase, PHB films were rapidly disintegrated and then completely degraded. At fourth phase, empty fibrous capsule resolved (Figure 1.5) [18, 31]. Interesting data were obtained for biodegradation *in vivo* of PHB microspheres (0.5–0.8 µm in diameter). It was demonstrated indirectly that PHB loss about 8 per cent of weight of microspheres accumulated in liver after 2 months of intravenous injection. A presence of several types PHB degrading enzymes in the animal tissues extracts [24] was demonstrated (Figure 1.6).

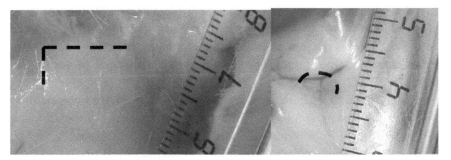

FIGURE 1.6 Biodegradation of PHB films *in Vivo*. Connective-tissue capsule with PHB thin films (outlined with broken line) 2 weeks (98% residual weight of the film) (left photograph) and 3 months (0% residual weight of the film) (left photograph) after subcutaneous implantation [18].

Some research works studied a biodegradation of PHB threads with a tendency of analysis of its mechanical properties *in vivo* [22, 23]. It was shown that at first load at break, index decreased rapidly from 440 to 390 g (12%) at 15th day and then gradually increased to the initial value at 90th and remain almost unchanged up to 182nd day [22] or gradually decreased in 27 per cent during 180 days [23], strain at break decreased rapidly from 60 to 50 per cent (in 17% of initial value) at 10th day and then gradually increased to 70 per cent (in 17% of initial value) at 182nd day [22] or did not change significantly during 180 days [23].

It was demonstrated that the primary reason of PHB biodegradation *in vivo* was a lysosomal and phagocytic activity of polynucleated macrophages and giant cells of foreign body reaction. The activity of tissue macrophages and nonspecific enzymes of body liquids made a main contribution to significantly more rapid rate of PHB biodegradation *in vivo* in comparison with rate of PHB hydrolysis *in vitro*. The PHB material was encapsulated by degrading macrophages. Presence of PHB stimulated uniform macrophage infiltration, which is important for not only the degradation process but also the restoration of functional tissue.

The long absorption time produced a foreign body reaction, which was restricted to macrophages forming a peripolymer layer [23, 65, 68, 72]. Very important data that clarifies the tissue response that contributes to biodegradation of PHB was obtained by Lobler. A significant increase of expression of two specific lipases after 7 and 14 days of PHB contact with animal tissues was demonstrated. Moreover, liver-specific genes were induced with similar results. It is striking that pancreatic enzymes are induced in the gastric wall after contact with biomaterials [46]. Saito et al. suggested the presence of at least two types of degradative enzymes in rat tissues: liver serine esterases with the maximum of activity in alkaline media (pH=9.5) and kidney esterases with the maximum of activity in neutral media [24].

The mechanism of PHB biodegradation by macrophages was demonstrated at cultured macrophages incubated with particles of low-molecular weight PHB [74]. It was shown that macrophages and, to a lesser level, fibroblasts have the ability to take up (phagocytize) PHB particles (1–10 μm). At high concentrations of PHB particles (>10 μg/mL), the phagocytosis is accompanied by toxic effects and alteration of the functional status of the macrophages but not the fibroblasts. This process is accompanied by cell damage and cell death.

The elevated production of nitric oxide (NO) and tumor necrosis factor alfa (TNF-α) by activated macrophages was observed. It was suggested that the cell damage and cell death may be due to phagocytosis of large amounts of PHB particles; after phagocytosis, polymer particles may fill up the cells, and cause cell damage and cell death. It was also demonstrated that phagocytized PHB particles disappeared in time because of an active PHB biodegradation process (Figure 1.6) [74].

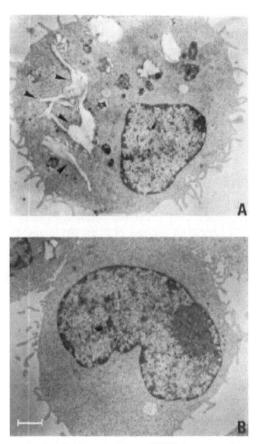

FIGURE 1.7 Phagocytosis of microparticles of PHB in macrophages. TEM analysis of cultured macrophages in the presence (A) or absence (B) of 2 μg PHB microparticles/mL for 24 h. Bar in B represents 1 μm, for A and B [74].

1.5 APPLICATION OF PHB

1.5.1 MEDICAL DEVICES ON THE BASE OF PHB AND PHB *IN VIVO* BIOCOMPATIBILITY

The perspective area of PHB application is the development of implanted medical devices for dental, cranio-maxillofacial, orthopedic, cardiovascular, hernioplastic, and skin surgery [3, 6]. A number of potential medical devices on the base of PHB: bioresorbable surgical sutures [6, 22–23, 31, 75–76], biodegradable screws and plates for cartilage and bone fixation [6, 24, 31, 62], biodegradable membranes for periodontal treatment [6, 31, 77, 78], surgical meshes with PHB coating for hernioplastic surgery [6, 31], wound coverings [79], patches for repair of a bowel, pericardial and osseous defects [14–15, 65–69], nerve guidance channels and conduits [20–21], cardiovascular stents [80], and so on was developed (Figure 1.7).

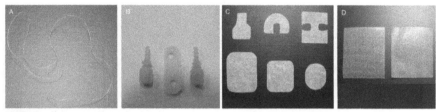

FIGURE 1.8 Medical devices on the base of PHB. (A) bioresorbable surgical suture; (B) biodegradable screws and plate for cartilage and bone fixation; (C) biodegradable membranes for periodontal treatment; (D) surgical meshes with PHB coating for hernioplastic surgery, pure (left) and loaded with antiplatelet drug, dipyridamole (right) [31].

The tissue reaction *in vivo* to implanted PHB films and medical devices was studied. In most cases, a good biocompatibility of PHB was demonstrated. In general, no acute inflammation, abscess formation, or tissue necrosis were observed in tissue surrounding of the implanted PHB materials. In addition, no tissue reactivity or cellular mobilization occurred in areas remote from the implantation site [13, 16, 31, 71]. On the one hand, it was shown that PHB elicited similar mild tissue response as PLA did [16], but on the other hand, the use of implants consisting of polylactic acid, polyglicolic acid, and their copolymers is not without a number of sequelae related with the chronic inflammatory reactions in tissue [81–85].

Subcutaneous implantation of PHB films for 1 month has shown that the samples were surrounded by a well-developed, homogeneous fibrous capsule of 80–100 μm in thickness. The vascularized capsule consists primarily of connective tissue cells (mainly, round, immature fibroblasts) aligned parallel to the implant surface. A mild inflammatory reaction was manifested by the presence of mononuclear macrophages, foreign body cells, and lymphocytes. Three months after implantation, the fibrous capsule has thickened to 180–200 μm because of the increase in the amount of connective tissue cells and a few collagen fiber deposits.

A substantial decrease in inflammatory cells was observed after 3 months, and tissues at the interface of the polymer were densely organized to form bundles. After 6 months of implantation, the number of inflammatory cells had decreased and the fibrous capsule, now thinned to about 80–100 μm, consisted mainly of collagen fibers, and a significantly reduced amount of connective tissue cells. A little inflammatory cells effusion was observed in the tissue adherent to the implants after 3 and 6 months of implantation [13, 16]. The biocompatibility of PHB has been demonstrated *in vivo* under subcutaneous implantation of PHB films. Tissue reaction to films from PHB of different molecular weights (300; 450; 1,000 kDa) implanted subcutaneously was relatively mild and did not change from tissue reaction to control glass plate [18, 31].

At implantation of PHB with contact to bone, the overall tissue response was favorable with a high rate of early healing and new bone formation with some indication of an osteogenic characteristic for PHB compared with other thermoplastics, such as polyethylene. Initially, there was a mixture of soft tissue, containing active fibroblasts, and rather loosely woven osteonal bone seen within 100 μm of the interface. There was no evidence of a giant cell response within the soft tissue in the early stages of implantation. With time this tissue became more orientated in the direction parallel to the implant interface.

The dependence of the bone growth on the polymer interface is demonstrated by the new bone growing away from the interface rather than toward it after implantation of 3 months. By 6 months postimplantation, the implant is closely encased in new bone of normal appearance with no interposed fibrous tissue. Thus, PHB-based materials produce superior bone healing [43].

Regeneration of a neointima and a neomedia, comparable to native arterial tissue, was observed at 3–24 months after implantation of PHB nonwoven patches as transannular patches into the right ventricular outflow tract and pulmonary artery. In the control group, a neointimal layer was present but no neomedia comparable to native arterial tissue. Three layers were identified in the regenerated tissue: neointima with a endothelium-like lining, neomedia with smooth muscle cells, collagenous and elastic tissue, and a layer with polynucleated macrophages surrounding istets of PHB, capillaries and collagen tissue. Lymphocytes were rare. It was concluded that PHB nonwoven patches can be used as a scaffold for tissue regeneration in low-pressure systems. The regenerated vessel had structural and biochemical qualities in common with the native pulmonary artery [68]. Biodegradable PHB patches implanted in atrial septal defects promoted formation of regenerated tissue that macroscopically and micro-scopically resembled native atrial septal wall.

The regenerated tissue was found to be composed of three layers: monolayer with endothelium-like cells, a layer with fibroblasts and some smooth-muscle cells, collagenous tissue and capillaries, and a third layer with phagocytizing cells isolating and degrading PHB. The neointima con-tained a complete endothelium-like layer resembling the native endothe-lial cells. The patch material was encapsulated by degrading macrophages. There was a strict border between the collagenous and the phagocytizing layer. Presence of PHB seems to stimulate uniform macrophage infiltra-tion, which was found to be important for the degradation process and the restoration of functional tissue.

Lymphocytic infiltration as foreign body reaction, which is common after replacement of vessel wall with commercial woven Dacron patch, was wholly absent when PHB. It was suggested that the absorption time of PHB patches was long enough to permit regeneration of a tissue with sufficient strength to prevent development of shunts in the atrial septal position [65]. The prevention of postoperative pericardial adhesions by closure of the pericardium with absorbable PHB patch was demonstrated. The regeneration of mesothelial layer after implantation of PHB pericar-dial patch was observed. The complete regeneration of mesothelium, with morphology and biochemical activity similar to findings in native meso-thelium, may explain the reduction of postoperative pericardial adhesions after operations with insertion of absorbable PHB patches [67].

The regeneration of normal filament structure of restored tissues was observed by immunohistochemical methods after PHB devices implan-tation [66]. The immunohistochemical demonstration of cytokeratine, an intermediate filament, which is a constituent of epithelial and mesodermal cells, agreed with observations on intact mesothelium. Heparin sulfate proteoglycan, a marker of basement membrane, was also identified [66]. However, in spite of good tissue reaction to implantation of cardiovascular PHB patches, PHB endovascular stents in the rabbit iliac arteria caused intensive inflammatory vascular reactions [80].

PHB patches for the gastrointestinal tract were tested using animal model. Patches made from PHB sutured and PHB membranes were implanted to close experimental defects of stomach and bowel wall. The complete regeneration of tissues of stomach and bowel wall was observed at 6 months after patch implantation without strong inflammatory response and fibrosis [14,86].

Recently, an application of biodegradable nerve guidance channels (conduits) for nerve repair procedures and nerve regeneration after spinal cord injury was demonstrated. Polymer tubular structures from PHB can be modulated for this purpose. Successful nerve regeneration through a guidance channel was observed as early as after 1 month. Virtually, all implanted conduits contained regenerated tissue cables centrally located within the channel lumen and composed of numerous myelinated axons and Schwann cells. The inflammatory reaction had not interfered with the nerve regeneration process. Progressive angiogenesis was present at the nerve ends and through the walls of the conduit. The results demonstrate good-quality nerve regeneration in PHB guidance channels [21, 87].

Biocompatibility of PHB was evaluated by implanting microspheres from PHB (M_w=450 kDa) into the femoral muscle of rats. The spheres were surrounded by one or two layers of spindle cells, and infiltration of inflammatory cells and mononuclear cells into these layers was recognized at 1 week after implantation. After 4 weeks, the number of inflammatory cells had decreased and the layers of spindle cells had thickened. No inflammatory cells were seen at 8 weeks, and the spheres were encapsulated by spindle cells. The toxicity of PHB microspheres was evaluated by weight change and survival times in L1210 tumor-bearing mice. No differences were observed in the weight change or survival time compared with those of control. These results suggest that inflammation accompanying microsphere implantation is temporary as well as toxicity to normal tissues is minimal [71].

The levels of tissue factors, inflammatory cytokines, and metabolites of arachidonic acid were evaluated. Growth factors derived from endothelium and from macrophages were found. These factors most probably stimulate both growth and regeneration occurring when different biodegradable materials were used as grafts [46, 65, 67, 86]. The positive reaction for thrombomodulin, a multifunctional protein with anticoagulant properties, was found in both mesothelial and endothelial cells after pericardial PHB patch implantation. Prostacycline production level, which was found to have cytoprotective effect on the pericardium and prevent adhesion formation, in the regenerated tissue was similar to that in native pericardium [65, 67]. The PHB patch seems to be highly biocompatible, as no signs of inflammation were observed macroscopically and also the level of inflammation associated cytokine mRNA did not change dramatically, although a transient increase of interleukin-1β and interleukin-6 mRNA through days 1-7 after PHB patch implantation was detected. In contrast, tumor necrosis factor-α mRNA was hardly detectable throughout the implantation period, which agrees well with a observed moderate fibrotic response [46, 86].

1.6   PHB AS TISSUE ENGINEERING MATERIAL AND PHB *IN VITRO* BIOCOMPATIBILITY

Biopolymer PHB is a promising material in tissue engineering because of its high biocompatibility *in vitro*. Cell cultures of various origins including murine and human fibroblasts [15, 40, 78, 88–92], human mesenchymal stem cells [93], rabbit bone marrow cells (osteoblasts) [36, 89, 94], human osteogenic sarcoma cells [95], human epithelial cells [90, 95], human endothelial cells [96–97], rabbit articular cartilage chondrocytes [98–99] and rabbit smooth muscle cells [100], and human neurons (Schwann cells) [101] in direct contact with PHB when cultured on polymer films and scaffolds exhibited satisfactory levels of cell adhesion, viability, and proliferation. Moreover, it was shown that fibroblasts, endothelium cells, and isolated hepatocytes cultured on PHB films exhibited high levels of cell adhesion and growth (Figure 1.8) [102]. A series of 2D and 3D PHB scaffolds was developed by various methods: polymer surface modification [97], blending [36, 78, 88, 90, 93, 99, 103], electrospinning [104–106], salt leaching [36, 94, 107, 108], microspheres fusion [109], forming on porous mold [110], and laser cutting [111].

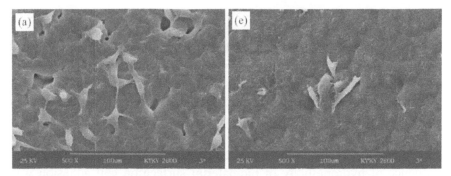

FIGURE 1.9   Scanning electron microscopy image of 2 days growth of fibroblast cells on films made of (a) PHB; (e) PLA; (500×). Cell density of fibroblasts grown on PHB film is significantly higher vs. cell density of fibroblasts grown on PLA film [89].

It was also shown that cultured cells produced collagen II and glycosaminoglycan, which are the specific structural biopolymers formed in the extracellular matrix [95, 98, 99]. A good viability and proliferation level of macrophages and fibroblasts cell lines were obtained in the culturing in the presence of particles from short-chain low-molecular PHB [74]. However, it was shown that cell growth on the PHB films was relatively poor: the viable cell number ranged from 1×10^3 to 2×10^5 [40, 89, 99]. An impaired interaction between PHB matrix and cytoskeleton of cultured cells was also demonstrated [95]. It was reported that a number of polymer properties including chemical composition, surface morphology, surface chemistry, surface energy, and hydrophobicity play important roles in regulating cell viability and growth [112]. The investigation showed that this biomaterial can be used to make scaffolds for *in vitro* proliferating cells [40, 89, 98].

The most widespread methods to manufacture the PHB scaffolds for tissue engineering by means of improvement of cell adhesion and growth on polymer surface are change of PHB surface properties and microstructure by salt-leaching methods and enzymatic/chemical/physical treatment of polymer surface [40, 89, 98, 113]. Adhesion to polymer substrates is one of the key issues in tissue engineering because adhesive interactions control cell physiology. One of the most effective techniques to improve adhesion and growth of cells on PHB films is the treatment of polymer surface with enzymes, alkali, or low-pressure plasma [40, 113]. Lipase treatment increases the viable cell number on the PHB film from 100 to 200 times compared to the untreated PHB film. NaOH treatment on PHB film also indicated an increase of 25 times on the viable cell number compared with the untreated PHB film [40].

It was shown that treatment of PHB film surface with low-pressure ammonia plasma improved the growth of human fibroblasts and epithelial cells of respiratory mucosa because of the increased hydrophylicity (but with no change of microstructure) of polymer surface [90]. It was suggested that the improved hydrophilicity of the films after PHB treatment with lipases, alkali, and plasma allowed cells in its suspension to easily attach on the polymer films compared to that on the untreated ones. The influence of hydrophilicity of biomaterial surface on cell adhesion was demonstrated earlier [114].

However, a microstructure of PHB film surface can also be responsible for cell adhesion and cell growth [115–117]. Therefore, above-noticed modification of polymer film surface after enzymatic and chemical treatments (in particular, reduced pore size and a surface smoothing) is expected to play an important role for enhanced cell growth on the polymer films [40]. Different cells prefer different surfaces. For example, osteoblasts preferred rougher surfaces with appropriate size of pores [115, 116] while fibroblast prefer smoother surface, yet epithelial cells only attached to the smoothest surface [117].

This appropriate roughness affects cell attachment as it provides the right space for osteoblast growth, or supplies solid anchors for filapodia. A scaffold with appropriate size of pores provided better surface properties for anchoring type II collagen filaments and for their penetration into internal layers of the scaffolds implanted with chondrocytes. This could be illuminated by the interaction of extracellular matrix proteins with the material surface. Moreover, the semicrystalline surface PHB ultrastructure can be connected with protein adsorption and cell adhesion [91, 92, 118]. The appropriate surface properties may also promote cell attachment and proliferation by providing more spaces for better gas/nutrients exchange or more serum protein adsorption [36, 94, 98]. Additionally Sevastianov et al. found that PHB films in contact with blood did not activate the hemostasis system at the level of cell response, but they did activate the coagulation system and the complement reaction [119].

The high biocompatibility of PHB may be due to several reasons. First of all, PHB is a natural biopolymer involved in important physiological functions in both prokaryotes and eukaryotes. PHB from bacterial origin has property of stereospecificity that is inherent to biomolecules of all living things and consists only from residues of D(-)-3-hydrohybutyric acid [120–122]. Low-molecular weight PHB (up to 150 resides of 3-hydrohybutyric acid), complexed to other macromolecules (cPHB), was found to be a ubiquitous constituent of both prokaryotic and eukaryotic organisms of nearly all phyla [123–127]. Complexed cPHB was found in a wide variety of tissues and organs of mammals (including human): blood, kidney, vessels, nerves, vessels, eye, brain, as well as in organelles, membrane proteins, lipoproteins, and plaques. cPHB concentration ranged from 3–4 µg/g wet tissue weight in nerves and brain to 12 µg/g in blood plasma [128, 129]. In humans, total plasma cPHB ranged from 0.60 to 18.2 mg/l, with a mean of 3.5 mg/l. [129]. It was shown that cPHB is a functional part of ion channels of erythrocyte plasma membrane and hepatocyte mitochondria membrane [130, 131].

The singular ability of cPHB to dissolve salts and facilitate their transfer across hydrophobic barriers defines a potential physiological niche for cPHB in cell metabolism [125]. However, a mechanism of PHB synthesis in eukaryotic organisms is not well clarified that requires additional studies. Nevertheless, it could be suggested that cPHB is one of the products of symbiotic interaction between animals and gut microorganisms. It was shown, for example, that *E. coli* is able to synthesize low-molecular weight PHB and cPHB plays various physiological roles in bacteria cells [127, 132].

Intermediate product of PHB biodegradation, D(-)-3-hydroxybutyric acid is also a normal constituent of blood at concentrations between 0.3 and 1.3 mM and is present in all animal tissues [133, 134]. As it was noted earlier PHB has a rather low degradation rate in the body in comparison to, for example, poly(lactic-*co*-glycolic) acids, that prevent from increase of 3-hydroxybutyric acid concentration in surrounding tissues [13, 16], whereas polylactic acid release, following local pH decrease in implantation area and acidic chronic irritation of surrounding tissues is a serious problem in application of medical devices on the base of poly(lactic-*co*-glycolic) acids [135, 136]. Moreover, chronic inflammatory response to polylactic and polyglycolic acids that was observed in a number of cases may be induced by immune response to water-soluble oligomers that released during degradation of synthetic polymers [136–138].

1.7   NOVEL DRUG DOSAGE FORMS ON THE BASE OF PHB

An improvement of medical materials on the base of biopolymers by encapsulating different drugs opens up the wide prospects in applications of new devices with pharmacological activity. The design of injection systems for sustained drug delivery in the forms of microspheres or microcapsules prepared on the base of biodegradable polymers is extremely challenging in the modern pharmacology.

The fixation of pharmacologically active component with the biopolymer and following slow drug release from the microparticles provides an optimal level of drug concentration in local target organ during long-term period (up to several months). At curative dose, the prolonged delivery of drugs from the systems into organism permits to eliminate the shortcomings in peroral, injectable, aerosol, and the other traditional methods of drug administration. Among those shortcomings hypertoxicity, instability, pulsative character of rate delivery, ineffective expenditure of drugs should be pointed out. Alternatively, applications of therapeutical polymer systems provide orderly and purposefully the deliverance for an optimal dose of agent that is very important at therapy of acute or chronic diseases [1, 3, 6, 139].

An ideal biodegradable microsphere formulation would consist of a free-flowing powder of uniform-sized microspheres less than 125 µm in diameter and with a high drug loading. In addition, the drug must be released in its active form with an optimized profile. The manufacturing method should produce the microspheres which are reproducible, scalable, and benign to some often delicate drugs, with a high encapsulation efficiency [140, 141].

PHB as biodegradable and biocompatible is a promising material for producing polymer systems for controlled drug release. A number of drugs with various pharmacological activities were used for the development of polymer-controlled release systems on the base of PHB and its copolymers: model drugs (2,7-dichlorofluorescein [142], dextran-FITC [143], methyl red [122, 144, 145], 7-hydroxethyltheophylline [146, 147]), antibiotics and antibacterial drugs (rifampicin [148, 149], tetracycline [150], cefoperazone and gentamicin [151], sulperazone and duocid [152–155], sulbactam and cefoperazone [156], fusidic acid [157], nitrofural [158]), anticancer drugs (5-fluorouracil [159], 2′,3′-diacyl-5-fluoro-2′-deoxyuridine [71], paclitaxel [160, 161], rubomycin [162], chlorambucil and etoposide [163]), antiinflammatory drug (indomethacin [164], flurbiprofen [165], ibuprofen [166]), analgesics (tramadol [167], vasodilator and antithrombotic drugs (dipyridamole [168, 169], NO donor [170, 171], nimodipine [174], felodipine [175]), proteins (hepatocyte growth factor [176], mycobacterial protein for vaccine development [177], and bone morphogenetic protein 7 [178]).

Various methods for manufacture of drug-loaded PHB matrices and microspheres were used: solvent casting of films [144–147], emulsification and solvent evaporation [148–164], spray drying [172], layer-by-layer self-assembly [165], and supercritical antisolvent precipitation [173]. The biocompatibility and pharmacological activity of some of these systems were studied [71, 148, 154–156, 160, 161, 167, 171]. However, only a few drugs were used for production of drug-controlled release systems on the base of PHB homopolymer: 7-hydroxethyltheophylline, methyl red, 2′,3′-diacyl-5-fluoro-2′-deoxyuridine, rifampicin, tramadol, indomethacin, dipyridamole, and paclitaxel [71, 154–156, 160–171]. The latest trend in PHB drug delivery systems study is the development of PHB nanoparticles loaded with different drugs [179–180].

The first drug-sustained delivery system on the base of PHB was developed by Korsatko et al., who observed a rapid release of encapsulated drug, 7-hydroxethyltheophylline, from tablets of PHB (M_w = 2,000 kDa), as well as weight losses of PHB tablets containing the drug after subcutaneous implantation. It was suggested that PHB with molecular weight greater than 100 kDa was undesirable for long-term medication dosage [146].

Pouton and Akhtar describing the release of low-molecular weight drugs from PHB matrices reported that the latter have the tendencies to enhance water penetration and pore formation [181]. The entrapment and release of the model drug, methyl red (MR), from melt-crystallized PHB was found to be a function of polymer crystallization kinetics and morphology whereas overall degree of crystallinity was shown to cause no effect on drug release kinetics. MR released from PHB films for 7 days with initial phase of rapid release ("burst effect") and second phase with relatively uniform release. Release profiles of PHB films crystallized at 110°C exhibited a greater burst effect when compared to those crystallized at 60°C. This was explained by better trapping of drug within polymeric spherulites with the more rapid rates of PHB crystallization at 110°C [144, 145].

Kawaguchi et al showed that chemical properties of drug and polymer molecular weight had a great impact on drug delivery kinetics from PHB matrix. Microspheres (100–300 μm in diameter) from PHB of different molecular weight (65,000, 135,000, and 450,000) were loaded with pro-drugs of 5-fluoro-2'-deoxyuridine (FdUR) synthesized by esterification with aliphatic acids (propionate, butyrate, and pentanoate). Prodrugs have different physicochemical properties, in particular, solubility in water (from 70 mg/ml for FdUR to 0.1 mg/ml for butyryl-FdUR). The release rates from the spheres depended on both the lipophilicity of the prodrug and the molecular weight of the polymer. Regardless of the polymer, the relative release rates were propionyle- FdUR > butyryl- FdUR > pentano-yl- FdUR. The release of butyryl- FdUR and pentanoyl- FdUR from the spheres consisting of low-molecular-weight polymer (M_w=65,000) was faster than that from the spheres of higher molecular weight (M_w=135,000 or 450,000). The effect of drug content on the release rate was also studied. The higher the drug content in the PHB microspheres, the faster was the drug release. The release of FdUR continued for more than 5 days [71].

Kassab developed a well-managed technique for manufacture of PHB microspheres loaded with drugs. Microspheres were obtained within a size of 5–100 μm using a solvent evaporation method by changing the initial polymer/solvent ratio, emulsifier concentration, stirring rate, and initial drug concentration. The drug overloading of up to 041 g rifampicin/g PHB was achieved. Drug release was rapid: the maximal duration of rifampicin delivery was 5 days. Both the size and drug content of PHB microspheres were found to be effective in controlling the drug release from polymer microspheres [149].

The sustained release of analgesic drug, tramadol, from PHB micro-spheres was demonstrated by Salman et al. It was shown that 58 per cent of the tramadol (the initial drug content in PHB matrix=18%) was released from the microspheres (7.5 μm in diameter) in the first 24 h. Drug release decreased with time. From 2 to 7 days, the drug release was with zero-order rate. The entire amount of tramadol was released after 7 days [167].

The kinetics of different drug releases from PHB micro- and nanopar-ticles loaded with dipyridamole, indomethacin, and paclitaxel was studied [160,161,164,168,169]. It was found that the release occurs via two mech-anisms, diffusion and degradation, operating simultaneously. Vasodilator and antithrombotic drug, dipyridamole, and anti-inflammatory drug, in-domethacin, diffusion processes determine the rate of the release at the early stages of the contact of the system with the environment (the first 6–8 days). The coefficient of the release diffusion of a drug depends on its nature, the thickness of the PHB films containing the drug, the weight ratio of dipyridamole and indomethacin in polymer, and the molecular weight of PHB. Thus, it is possible to regulate the rate of drug release by changing of molecular weight of PHB, for example [164]. The biodegradable microspheres on the base of PHB designed for controlled release of dipyridamole and paclitaxel were kinetically studied. The profiles of release from the microspheres with different diameters present the progression of nonlinear and linear stages. Diffusion kinetic equation describing both linear (PHB hydrolysis) and nonlinear (diffusion) stages of the dipyridamole and paclitaxel release profiles from the spherical subjects has been written down as the sum of two terms: desorption from the homogeneous sphere in accordance with diffusion mechanism and the zero-order release.

In contrast to the diffusivity dependence on microsphere size, the constant characteristics of linearity are scarcely affected by the diameter of PHB microparticles. The view of the kinetic profiles as well as the low rate of dipyridamole and paclitaxel release are in satisfactory agreement with kinetics of weight loss measured *in vitro* for the PHB films and observed qualitatively for PHB microspheres. Taking into account kinetic results, it was supposed that the degradation of PHB microspheres is responsible for the linear stage of dipyridamole and paclitaxel release profiles (Figures 1.9 and 1.10) [24, 160, 161, 168, 169].

The biocompatibility and pharmacological activity of advanced drug delivery systems on the base of PHB were studied [71, 148, 167–169]. It was shown that implanted PHB microspheres loaded with paclitaxel caused the mild tissue reaction. The inflammation accompanying implantation of PHB matrices is temporary and additionally toxicity relative to normal tissues is minimal [169]. No signs of toxicity were observed after administration of PHB microspheres loaded with analgesic, tramadol [167]. A single intraperitoneal injection of PHB microspheres containing anticancer prodrugs, butyryl-FdUR, and pentanoyl-FdUR, resulted in high antitumor effects against P388 leukemia in mice over a period of 5 days [71]. Embolization with PHB microspheres *in vivo* at dogs as test animals has been studied by Kasab et al. Renal angiograms obtained before and after embolization and also the histopathological observations showed the feasibility of using these microspheres as an alternative chemoembolization agent [148]. Epidural analgesic effects of tramadol released from PHB

microspheres were observed for 21 h, whereas an equal dose of free tramadol was effective for less than 5 h. It was suggested that controlled release
of tramadol from PHB microspheres *in vivo* is possible, and pain relief
during epidural analgesia is prolonged by this drug formulation compared
with free tramadol [167].

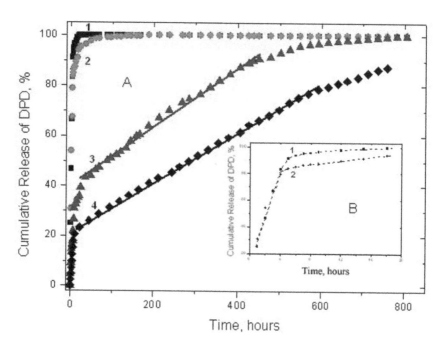

FIGURE 1.10 Kinetics profiles of DPD release from PHB microspheres *in vitro*
(phosphate buffer, 37°C). A: general view of kinetic curves for the microspheres with
different diameters: 4(1), 19(2), 63(3), and 92(4) μm. The lines show the second stage of
release following the zero-order equation. B: Details of the curves for the microspheres
with the smaller diameters: 4(1), 19(2).

The observed data indicate the wide prospects in applications of drug-
loaded medical devices and microspheres on the base of PHB as implant-
able and injectable therapeutic systems in medicine for treatment of various diseases: cancer, cardiovascular diseases, tuberculosis, osteomyelitis,
arthritis, and so on [6].

1.8   CONCLUSIONS

The natural PHB is unique biodegradable thermoplastics of considerable commercial importance. With this review, we have attempted to systematically evaluate the impact of physicochemical factors on the hydrolysis and the biodegradation of natural PHB both *in vitro* and *in vivo*. Clearly, the degradation behavior observed is very much dependent on both physicochemical conditions, geometry, and structural and microbial properties. If these conditions of (bio)degradation are known, the systems on the base PHB can be designed in such biomedicine areas as medical devices (Section 3.1), tissue scaffolds in bioengineering (Section 3.2), and development of novel biodegradable therapeutic systems for drug delivery.

ACKNOWLEDGMENTS

The work was supported by the Russian Foundation for Basic Research (grant no. 14-03-00405-a) and the Russian Academy of Sciences under the program "Construction of New_Generation Macromolecular Structures" (03/OC-14).

KEYWORDS

- **Biodegradable polymers**
- **Enzymatic degradation**
- **Hydrolysis**
- **PHB**

REFERENCES

1. Rice, J. J.; Martino, M. M.; De Laporte, L.; Tortelli, F.; Briquez, P. S.; and Hubbell, J. A.; *Adv. Healthcare Mater.* **2013**, *2*, 57–71. DOI: 10.1002/adhm.201200197. *Eng. Regenerat. Microenviron. Biomater.*
2. Khademhosseini, A.; and Peppas, N. A.; *Adv. Healthcare Mater.* **2013**, *2*, 10–12 DOI: 10.1002/adhm.201200444. *Micro and Nanoeng. Biomater. Healthcare Appl.*
3. Chen, G. Q.; Wu, Q. *Biomater.* **2005**, *26(33)*, 6565–6578. The application of polyhydroxyalkanoates as tissue engineering materials.

4. Lenz, R. W.; and Marchessault, R. H.; *Biomacromole.* **2005,** *6(1),* 1–8. *Bacterial Polyesters: Biosynthesis, Biodegradable Plastics and Biotechnol.*

5. Anderson, A. J.; and Dawes, E. A.; *Microbiol. Rev.* **1990,** *54(4),* 450–472. *Occurrence, Metabolism, Metabolic Role, Ind. Uses Bacterial Polyhydroxyalkanoates.*

6. Bonartsev, A. P.; Bonartseva, G. A.; Shaitan, K. V.; and Kirpichnikov M. P.; *Biochemistry (Moscow) Supp. Ser. B: Biomed. Chem.* **2011,** *5(1),* 10–21. *Poly(3-Hydroxybutyrate) and Poly(3-Hydroxybutyrate)-Based Biopolym. Systems.*

7. Jendrossek, D.; and Handrick, R.; *Annu Rev Microbiol.* **2002,** *56,* 403–432. *Microbial degradation of polyhydroxyalkanoates.*

8. Fabra M. J.; Lopez-Rubio A.; Lagaron J. M.; Food Hydrocolloids **2013,** *32*: 106–114. DOI http://dx.doi.org/10.1016/j.foodhyd.2012.12.007 High barrier polyhydroxyalcanoate food packaging film by means of nanostructured electrospun interlayers of zein.

9. Kim, D. Y.; and Rhee, Y. H.; *Appl. Microbiol. Biotechnol.* **2003,** *61,* 300–308. *Biodegradation of Microbial and Synthetic Polyesters by Fungi.*

10. Marois, Y.; Zhang, Z.; Vert, M.; Deng, X.; Lenz, R.; and Guidoin, R. J.; *Biomater. Sci. Polym. Ed.* **1999,** *10,* 483–499. *Hydrolytic and Enzymatic Incubation of Polyhydroxyoctanoate (PHO): A Short-Term in Vitro Study of a Degradable Bacterial Polyester.*

11. Abe, H.; and Doi, Y.; *Biomacromole.* **2002,** *3(1),* 133–138. *Side-Chain Effect of Second Monomer Units on Crystalline Morphology, Thermal Properties, and Enzymatic Degradability for Random Copolyesters of (R)-3-Hydroxybutyric Acid with (R)-3-Hydroxyalkanoic Acids.*

12. Renstad, R.; Karlsson, S.; and Albertsson, A. C.; *Polym. Degrad. Stab.* **1999,** *63,* 201–211. *The Influence of Processing Induced Differences in Molecular Structure on the Biological and Non-Biological Degradation of Poly(3-Hydroxybutyrate-co-3-Hydroxyvalerate), P(3-HB-co-3-HV).*

13. Qu, X. H.; Wu, Q., Zhang, K. Y.; and Chen, G. Q.; *Biomater.* **2006,** *27(19),* 3540–3548. *In Vivo Studies of Poly(3-Hydroxybutyrate-co-3-Hydroxyhexanoate) Based Polymers: Biodegradation and Tissue Reactions.*

14. Freier, T.; et al. *Biomater.* **2002,** *23(13),* 2649–2657. *In Vitro and in Vivo Degradation Studies for Development of a Biodegradable Patch Based on Poly(3-Hydroxybutyrate).*

15. Kunze, C.; et al. *Biomater.* **2006** January, *27(2),* 192–201. *In Vitro and in Vivo Studies on Blends of Isotactic and Atactic Poly (3-Hydroxybutyrate) for Development of a Dura Substitute Material.*

16. Gogolewski, S.; Jovanovic, M.; Perren, S. M.; Dillon, J. G.; and Hughes, M. K. *J.; Biomed. Mater. Res.* **1993,** *27(9),* 1135–1148. *Tissue response and in vivo degradation of selected polyhydroxyacids: polylactides (PLA), poly(3-hydroxybutyrate) (PHB), and poly(3-hydroxybutyrate-co-3-hydroxyvalerate) (PHB/VA).*

17. Boskhomdzhiev, A. P.; et al. *Int. Polym. Sci. Technol.* **2010,** *37(11),* 25–30. *Hydrolytic degradation of biopolymer systems based on poly(3-hydroxybutyrate. Kinetic and structural aspects.*

18. Boskhomdzhiev, A. P.; et al. *Biochem. (Moscow) Supp. Ser. B: Biomed. Chem.* **2010,** *4(2),* 177–183. *Biodegradation kinetics of poly(3-hydroxybutyrate)-based biopolymer systems.*

19. Borkenhagen, M.; Stoll, R. C.; Neuenschwander, P.; Suter, U. W.; and Aebischer, P.; Biomaterials **1998**, *19(23)*, 2155–2165. *In vivo performance of a new biodegradable polyester urethane system used as a nerve guidance channel.*

20. Hazari, A.; et al. *J. Hand Surgery (British and Euro. Vol.* **1999**) *24B(3)*, 291–295. *A new resorbable wrap-around implant as an alternative nerve repair technique.*

21. Hazari, A.; Wiberg, M.; Johansson-Rudén, G.; Green, C.; and Terenghi, G. A.; *British J. Plastic Surgery.* **1999**, *52*, 653–657. *Resorbable nerve conduit as an alternative to nerve autograft.*

22. Miller, N. D.; and Williams, D. F.; *Biomater.* **1987** March, *8(2)*, 129–137. *On the biodegradation of poly-beta-hydroxybutyrate (PHB) homopolymer and poly-beta-hydroxybutyrate-hydroxyvalerate copolymers.*

23. Shishatskaya, E. I.; Volova, T. G.; Gordeev, S. A.; and Puzyr, A. P. J.; *Biomater Sci Polym Ed.* **2005**, *16(5)*, 643–657. *Degradation of P(3HB) and P(3HB-co-3HV) in biological media.*

24. Bonartsev, A. P.; Livshits, V. A.; Makhina, T. A.; Myshkina, V. L.; Bonartseva, G. A.; and Iordanskii1, A. L.; *Express Polym. Lett.* **2007**, *1(12)*, 797–803. DOI: 10.3144/expresspolymlett.2007.110 *Controlled release profiles of dipyridamole from biodegradable microspheres on the base of poly(3-hydroxybutyrate).*

25. Koyama, N.; and Doi, Y.; Morphology and biodegradability of a binary blend of poly((R)-3-hydroxybutyric acid) and poly((R,S)-lactic acid). Can. *J. Microbiol.* **1995**, *41*(Suppl. 1), 316–322.

26. Doi, Y.; Kanesawa, Y.; Kunioka, M.; and Saito, T.; Biodegradation of microbial copolyesters: poly(3-hydroxybutyrate-co-3-hydroxyvalerate) and poly(3-hydroxybutyrate-co-4- hydroxybutyrate). *Macromole.* **1990a**, *23*, 26–31.

27. Holland, S.J.; Jolly, A. M.; Yasin, M.; and Tighe, B. J.; Polymers for biodegradable medical devices. II. Hydroxybutyrate-hydroxyvalerate copolymers: hydrolytic degradation studies. *Biomater.* **1987**, *8(4)*, 289–295.

28. Kurcok, P.; Kowalczuk, M.; Adamus, G.; Jedlinrski, Z.; and Lenz, R. W.; Degradability of poly(b-hydroxybutyrate)s. Correlation with chemical microstucture. *JMS-Pure Appl. Chem.* **1995**, *A32*, 875–880.

29. Bonartsev, A. P.; et al. Hydrolytic degradation of poly(3-hydroxybutyrate), polylactide and their derivatives: kinetics, crystallinity, and surface morphology. *Mole. Cryst. Liquid Cryst.* **2012**, *556(1)*, 288–300.

30. Bonartsev, A. P.; et al. Hydrolytic degradation of poly(3-hydroxybutyrate) and its derivates: characterization and kinetic behavior. *Chem. Chem. Technol.* **2012, T.6, N.4**, 385–392.

31. Bonartsev, A. P.; et al. Biosynthesis, biodegradation, and application of poly(3-hydroxybutyrate) and its copolymers natural polyesters produced by diazotrophic bacteria. *Comm. Current Res. Educat. Topics Trends Appl. Microbiol.* Ed. Méndez-Vilas, A.; Formatex, Spain; **2007**, *1*, 295–307.

32. Cha, Y.; and Pitt, C. G.; The biodegradability of polyester blends. *Biomater.* **1990**, *11(2)*, 108–112.

33. Schliecker, G.; Schmidt, C.; Fuchs, S.; Wombacher, R.; and Kissel, T.; Hydrolytic degradation of poly(lactide-co-glycolide) films: effect of oligomers on degradation rate and crystallinity. *Int. J. Pharm.* **2003**, *266(1–2)*, 39–49.

34. Scandola, M.; et al. Polymer blends of natural poly(3-hydroxybutyrate-co-hydroxyval-erate) and a synthetic atactic poly(3-hydroxybutyrate). Characterization and biodegra-dation studies. *Macromole.* **1997,** *30,* 2568–2574.

35. Doi, Y.; Kanesawa, Y.; Kawaguchi, Y.; and Kunioka, M.; Hydrolytic degradation of microbial poly(hydroxyalkanoates). *Makrom. Chem. Rapid. Commun.* **1989,** *10,* 227–230.

36. Wang, Y. W.; et al. Evaluation of three-dimensional scaffolds made of blends of hy-droxyapatite and poly(3-hydroxybutyrate-co-3-hydroxyhexanoate) for bone recon-struction *Biomater.* **2005,** *26(8),* 899–904 (a).

37. Muhamad, I. I.; Joon, L. K.; Noor, M. A. M.; Comparing the degradation of poly-β-(hydroxybutyrate), poly-β-(hydroxybutyrate-co-valerate) (PHBV) and PHBV/Cellu-lose triacetate blend. *Malaysian Polym. J.* **2006,** *1,* 39–46.

38. Mergaert, J.; Webb, A.; Anderson, C.; Wouters, A.; and Swings, J.; Microbial degrada-tion of poly(3-hydroxybutyrate) and poly(3-hydroxybutyrate-co-3-hydroxyvalerate) in soils. *Appl. Environ. Microbiol.* **1993,** *59(10),* 3233–3238.

39. Choi, G. G.; Kim, H. W.; and Rhee, Y. H.; Enzymatic and non-enzymatic degradation of poly(3-hydroxybutyrate-co-3-hydroxyvalerate) copolyesters produced by Alcalig-enes sp. MT-16. *J. Microbiol.* **2004,** *42(4),* 346–352.

40. Yang, X.; Zhao, K.; and Chen, G. Q.; Effect of surface treatment on the biocompat-ibility of microbial polyhydroxyalkanoates. *Biomater.* **2002,** *23(5),* 1391–1397.

41. Zhao, K.; Yang, X.; Chen, G. Q.; and Chen, J. C.; Effect of lipase treatment on the bio-compatibility of microbial polyhydroxyalkanoates. *J. Mater. Sci.: Mater. Med.* **2002,** *13,* 849–854.

42. Wang, H. T.; Palmer, H.; Linhardt, R. J.; Flanagan, D. R.; and Schmitt, E.; Degradation of poly(ester) microspheres. *Biomater.* **1990,** *11(9),* 679–685.

43. Doyle, C.; Tanner, E. T.; and Bonfield, W.; In vitro and in vivo evaluation of polyhy-droxybutyrate and of polyhydroxybutyrate reinforced with hydroxyapatite. *Biomater.* **1991,** *12,* 841–847.

44. Coskun, S.; Korkusuz, F.; and Hasirci, V.; Hydroxyapatite reinforced poly(3-hydroxy-butyrate) and poly(3-hydroxybutyrate-co-3-hydroxyvalerate) based degradable com-posite bone plate. *J. Biomater. Sci. Polym. Edn.* **2005,** *16(12),* 1485–1502.

45. Sudesh, K.; Abe, H.; and Doi, Y.; Synthesis, structure and properties of polyhydroxy-alkanoates: biological polyesters. *Prog. Polym. Sci.* **2000,** *25,* 1503–1555

46. Lobler, M.; Sass, M.; Kunze, C.; Schmitz, K. P.; and Hopt, U. T.; Biomaterial patches sutured onto the rat stomach induce a set of genes encoding pancreatic enzymes. *Bio-mater.* **2002,** *23,* 577–583.

47. Bagrov, D. V.; et al. Amorphous and semicrystalline phases in ultrathin films of poly(3-hydroxybutirate). TechConnect World NTSI-Nanotech **2012** Proceedings, ISBN 978-1-4665-6274-5, **2012,** *1,* 602–605.

48. Kikkawa, Y.; Suzuki, T.; Kanesato, M.; Doi, Y.; and Abe, H.; Effect of phase structure on enzymatic degradation in poly(L-lactide)/atactic poly(3-hydroxybutyrate) blends with different miscibility. *Biomacromole.* **2009,** *10(4),* 1013–1018.

49. Tokiwa, Y.; Suzuki, T.; and Takeda, K.; Hydrolysis of polyesters by Rhizopus arrhizus lipase. *Agric. Biol. Chem.* **1986,** *50,* 1323–1325.

50. Hoshino, A.; and Isono, Y.; Degradation of aliphatic polyester films by commercially available lipases with special reference to rapid and complete degradation of poly(L-

lactide) film by lipase PL derived from Alcaligenes sp. Biodegradation **2002**, *13*, 141–147.

51. Misra, S. K.; et al. Effect of nanoparticulate bioactive glass particles on bioactivity and cytocompatibility of poly(3-hydroxybutyrate) composites. *J. Res. Soc. Interface.* **2010**, *7(44)*, 453–465.

52. Jendrossek, D.; Schirmer, A.; and Schlegel, H. G.; Biodegradation of polyhydroxyalkanoic acids. Appl. Microbiol. *Biotechnol.* **1996**; *46*, 451–463.

53. Winkler, F. K.; D'Arcy, A.; and Hunziker, W.; Structure of human pancreatic lipase. *Nature.* **1990**; *343*, 771–774.

54. Jesudason, J. J.; Marchessault, R. H.; and Saito, T.; Enzymatic degradation of poly([R,S] β-hydroxybutyrate). *J. Environ. Polym. Degradat.* **1993**, *1(2)*, 89–98.

55. Mergaert, J.; Anderson, C.; Wouters, A.; Swings, J.; and Kersters, K.; Biodegradation of polyhydroxyalkanoates. *FEMS Microbiol. Rev.* **1992**, *9(2–4)*, 317–321.

56. Tokiwa, Y.; and Calabia, B. P.; Degradation of microbial polyesters. *Biotechnol. Lett.* **2004**, *26(15)*, 1181–1189.

57. Mokeeva, V.; Chekunova, L.; Myshkina, V.; Nikolaeva, D.; Gerasin, V.; and Bonartseva G. Biodestruction of poly(3-hydroxybutyrate) by microscopic fungi: tests of polymer on resistance to fungi and fungicidal properties. *Mikologia and Fitopatologia.* **2002**, *36(5)*, 59–63.

58. Bonartseva, G. A.; et al. Aerobic and anaerobic microbial degradation of poly-beta-hydroxybutyrate produced by Azotobacter chroococcum. *Appl. Biochem. Biotechnol.* **2003**, *109(1–3)*, 285–301.

59. Spyros, A.; Kimmich, R.; Briese, B. H.; and Jendrossek, D.; 1H NMR Imaging Study of Enzymatic Degradation in Poly(3-hydroxybutyrate) and Poly(3-hydroxybutyrate-co-3-hydroxyvalerate). Evidence for Preferential Degradation of the Amorphous Phase by PHB Depolymerase B from Pseudomonas lemoignei. *Macromole.* **1997**, *30(26)*, 8218–8225.

60. Hocking, P. J., Marchessault, R. H.; Timmins, M. R.; Lenz, R. W.; and Fuller, R. C.; Enzymatic degradation of single crystals of bacterial and synthetic poly(-hydroxybutyrate) *Macromole.* **1996**, *29(7)*, 2472–2478.

61. Bonartseva, G. A.; Myshkina, V. L.; Nikolaeva, D. A.; Rebrov, A. V.; Gerasin, V. A.; and Makhina, T. K.; [The biodegradation of poly-beta-hydroxybutyrate (PHB) by a model soil community: the effect of cultivation conditions on the degradation rate and the physicochemical characteristics of PHB]. *Mikrobiologiia.* **2002**, *71(2)*, 258–263, Russian.

62. Kramp, B.; et al. [Poly-beta-hydroxybutyric acid (PHB) films and plates in defect covering of the osseus skull in a rabbit model]. *Laryngorhinootologie.* **2002**, *81(5)*, 351–356, [Article in German].

63. Misra, S. K.; et al. Poly(3-hydroxybutyrate) multifunctional composite scaffolds for tissue engineering applications. *Biomater.* **2010**, *31(10)*, 2806–2815.

64. Kuppan, P.; Vasanthan, K. S.; Sundaramurthi, D.; Krishnan, U. M.; and Sethuraman, S.; Development of poly(3-hydroxybutyrate-co-3-hydroxyvalerate) fibers for skin tissue engineering: effects of topography, mechanical, and chemical stimuli. *Biomacromole.* **2011**, *12(9)*, 3156–3165.

65. Malm, T.; Bowald, S.; Karacagil, S.; Bylock, A.; and Busch, C.; A new biodegradable patch for closure of atrial septal defect. An experimental study. *Scand J Thorac Cardiovasc Surg.* **1992**; *26(1)*, 9–14(a).

66. Malm, T.; Bowald, S.; Bylock, A.; Saldeen, T.; and Busch, C.; Regeneration of pericardial tissue on absorbable polymer patches implanted into the pericardial sac. An immunohistochemical, ultrastructural and biochemical study in the sheep. *Scandinavian J. Thoracic Cardiovascular Surg.* **1992**, *26(1)*, 15–21(b).

67. Malm, T.; Bowald, S.; Bylock, A.; and Busch, C.; Prevention of postoperative pericardial adhesions by closure of the pericardium with absorbable polymer patches. An experimental study. *J. Thoracic Cardiovascular Surg.* **1992**, *104*, 600–607(c).

68. Malm, T.; Bowald, S.; Bylock, A.; Busch, C.; and Saldeen, T.; Enlargement of the right ventricular outflow tract and the pulmonary artery with a new biodegradable patch in transannular position. European Surgical Research, **1994**, *26*, 298–308.

69. Duvernoy, O.; Malm, T.; Ramström, J.; and Bowald, S.; A biodegradable patch used as a pericardial substitute after cardiac surgery: 6 and 24-month evaluation with CT. *Thorac Cardiovasc Surg.* **1995** October, *43(5)*, 271–274.

70. Unverdorben, M.; et al. A polyhydroxybutyrate biodegradable stent: preliminary experience in the rabbit. *Cardiovasc Intervent Radiol.* **2002**, *25(2)*, 127–132.

71. Kawaguchi, T.; et al. Control of drug release with a combination of prodrug and polymer matrix: antitumor activity and release profiles of 2',3'-Diacyl-5-fluoro-2'-deoxyuridine from poly(3-hydroxybutyrate) microspheres. *J. Pharmaceutical Sci.* **1992**, *87(6)*, 508–512.

72. Baptist, J. N.; (Assignor to W.R. Grace Et Co., New York), US Patent No. 3 *225 766*, **1965**

73. Holmes, P.; Biologically produced (R)-3-hydroxy-alkanoate polymers and copolymers. In: Ed. Bassett, D. C.; Developments in Crystalline Polymers. London: Elsevier; **1988**, *2*, 1–65.

74. Saad, B.; et al. Characterization of the cell response of cultured macrophages and fibroblasts td particles of short-chain poly[(R)-3-hydroxybutyric acid]. *J. Biomed. Mater. Res.* **1996**, *30*, 429–439.

75. Fedorov, M.; Vikhoreva, G.; Kildeeva, N.; Maslikova, A.; Bonartseva, G.; and Galbraikh, L.; [Modeling of surface modification process of surgical suture]. *Chimicheskie Volokna.* **2005**, *(6)*, 22–28, [Article in Russian].

76. Rebrov, A. V.; Dubinskii, V. A.; Nekrasov, Y. P.; Bonartseva, G. A.; Shtamm, M.; and Antipov, E. M.; [Structure phenomena at elastic deformation of highly oriented polyhydroxybutyrate]. *Vysokomol. Soedin. (Russian)* **2002**, *44*, 347–351. [Article in Russian].

77. Kostopoulos, L.; and Karring, T.; Augmentation of the rat mandible using guided tissue regeneration. *Clin Oral Implants Res.* **1994**, *5(2)*, 75–82.

78. Zharkova, I. I.; et al. The effect of poly(3-hydroxybutyrate) modification by poly(ethylene glycol) on the viability of cells grown on the polymer films]. *Biomed. Khim.* **2012**, *58(5)*, 579–591.

79. Kil'deeva, N. R.; et al. [Preparation of biodegradable porous films for use as wound coverings] *Prikl. Biokhim. Mikrobiol.* **2006**, *42(6)*, 716–720, [Article in Russian].

80. Unverdorben, M.; et al. Polyhydroxybutyrate biodegradable stent: preliminary experience in the rabbit. *Cardiovasc. Intervent. Radiol.* **2002**, *25*, 127–132.

81. Solheim, E.; Sudmann, B.; Bang, G.; and Sudmann, E.; Biocompatibility and effect on osteogenesis of poly(ortho ester) compared to poly(DL-lactic acid). *J. Biomed. Mater. Res.* **2000**, *49(2)*, 257–263.

82. Bostman, O.; and Pihlajamaki, H.; Clinical biocompatibility of biodegradable orthopaedic implants for internal fixation: a review. *Biomater.* **2000**, *21(24)*, 2615–2621.

83. Lickorish, D.; Chan, J.; Song, J.; and Davies, J. E.; An in-vivo model to interrogate the transition from acute to chronic inflammation. *Eur. Cell. Mater.* **2004**, *8*, 12–19.

84. Khouw, I. M.; van Wachem, P. B.; de Leij, L. F.; and van Luyn, M. J.; Inhibition of the tissue reaction to a biodegradable biomaterial by monoclonal antibodies to IFN-gamma. *J. Biomed. Mater. Res.* **1998**, *41*, 202–210.

85. Su, S. H.; Nguyen, K. T.; Satasiya, P.; Greilich, P. E.; Tang, L.; and Eberhart, R. C.; Curcumin impregnation improves the mechanical properties and reduces the inflammatory response associated with poly(L-lactic acid) fiber. *J. Biomater. Sci. Polym. Ed.* **2005**, *16(3)*, 353–370.

86. Lobler, M.; Sass, M.; Schmitz, K. P.; and Hopt, U. T.; Biomaterial implants induce the inflammation marker CRP at the site of implantation. *J. Biomed. Mater. Res.* **2003**, *61*, 165–167.

87. Novikov, L. N.; Novikova, L. N.; Mosahebi, A.; Wiberg, M.; Terenghi, G.; and Kellerth, J. O.; A novel biodegradable implant for neuronal rescue and regeneration after spinal cord injury. *Biomater.* **2002**, *23*, 3369–3376.

88. Cao, W.; Wang, A.; Jing, D.; Gong, Y.; Zhao, N.; and Zhang, X.; Novel biodegradable films and scaffolds of chitosan blended with poly(3-hydroxybutyrate). *J. Biomater. Sci. Polym. Edn.* **2005**, *16(11)*, 1379–1394.

89. Wang, Y. W.; et al. Effect of composition of poly(3-hydroxybutyrate-co-3-hydroxyhexanoate) on growth of fibroblast and osteoblast. *Biomater.* **2005**, *26(7)*, 755–761(b).

90. Ostwald, J.; Dommerich, S.; Nischan, C.; and Kramp, B.; [In vitro culture of cells from respiratory mucosa on foils of collagen, poly-L-lactide (PLLA) and poly-3-hydroxybutyrate (PHB)]. *Laryngorhinootol.* **2003**, *82(10)*, 693–699, [Article in Germany].

91. Bonartsev, A. P.; et al. The terpolymer produced by Azotobacter chroococcum 7B: effect of surface properties on cell attachment. *Plos One.* *8(2)*, e57200.

92. Bonartsev, A. P.; et al. Cell attachment on poly(3-hydroxybutyrate)-poly(ethylene glycol) copolymer produced by Azotobacter chroococcum 7B. *BMC Biochem.* **2013**, (in press).

93. Wollenweber, M.; Domaschke, H.; Hanke, T.; Boxberger, S.; Schmack, G.; Gliesche, K.; Scharnweber, D.; Worch, H.; Mimicked bioartificial matrix containing chondroitin sulphate on a textile scaffold of poly(3-hydroxybutyrate) alters the differentiation of adult human mesenchymal stem cells. *Tissue Eng.* **2006** February, *12(2)*, 345–359.

94. Wang, Y. W.; Wu, Q.; and Chen, G. Q.; Attachment, proliferation and differentiation of osteoblasts on random biopolyester poly(3-hydroxybutyrate-co-3-hydroxyhexanoate) scaffolds. *Biomater.* **2004**, *25(4)*, 669–675.

95. Nebe, B.; et al. Structural alterations of adhesion mediating components in cells cultured on poly-beta-hydroxy butyric acid. *Biomater.* **2001**, *22(17)*, 2425–2434.

96. Qu, X. H.; Wu, Q.; and Chen, G. Q.; In vitro study on hemocompatibility and cytocompatibility of poly(3-hydroxybutyrate-co-3-hydroxyhexanoate). *J. Biomater. Sci. Polymer Edn.* **2006**, *17(10)*, 1107–1121(a).

97. Pompe, T.; Keller, K.; Mothes, G.; Nitschke, M.; Teese, M.; Zimmermann, R.; Werner, C.; Surface modification of poly(hydroxybutyrate) films to control cell-matrix adhesion. *Biomater.* **2007,** *28(1),* 28–37.

98. Deng, Y.; et al. Poly(hydroxybutyrate-co-hydroxyhexanoate) promoted production of extracellular matrix of articular cartilage chondrocytes in vitro. *Biomater.* **2003,** *24(23),* 4273–4281.

99. Zheng, Z.; Bei, F. F.; Tian, H. L.; and Chen, G. Q.; Effects of crystallization of polyhydroxyalkanoate blend on surface physicochemical properties and interactions with rabbit articular cartilage chondrocytes, *Biomater.* **2005,** *26,* 3537–3548.

100. Qu, X. H.; Wu, Q.; Liang, J.; Zou, B.; and Chen, G. Q.; Effect of 3-hydroxyhexanoate content in poly(3-hydroxybutyrate-co-3-hydroxyhexanoate) on in vitro growth and differentiation of smooth muscle cells. *Biomater.* **2006** May, *27(15),* 2944–2950.

101. Sangsanoh, P.; et al. In vitro biocompatibility of schwann cells on surfaces of biocompatible polymeric electrospun fibrous and solution-cast film scaffolds. *Biomacromole.* **2007,** *8(5),* 1587–1594.

102. Shishatskaya, E. I.; and Volova, T. G.; A comparative investigation of biodegradable polyhydroxyalkanoate films as matrices for in vitro cell cultures. *J. Mater. Sci-Mater. M.* **2004,** *15,* 915–923.

103. Iordanskii, A. L.; et al. Hydrophilicity impact upon physical properties of the environmentally friendly poly(3-hydroxybutyrate) blends: modification via blending. Fillers, Filled Polymers and Polymer Blends, Willey-VCH; **2006** г, **233,** 108–116.

104. Suwantong, O.; Waleetorncheepsawat, S.; Sanchavanakit, N.; Pavasant, P.; Cheepsunthorn, P.; Bunaprasert, T.; Supaphol, P.; In vitro biocompatibility of electrospun poly(3-hydroxybutyrate) and poly(3-hydroxybutyrate-co-3-hydroxyvalerate) fiber mats. *Int. J. Biol Macromol.* **2007,** *40(3),* 217–23.

105. Heidarkhan Tehrani, A.; Zadhoush, A.; Karbasi, S.; and Sadeghi-Aliabadi, H.; Scaffold percolative efficiency: in vitro evaluation of the structural criterion for electrospun mats. *J. Mater. Sci. Mater. Med.* **2010,** *21(11),* 2989–2998.

106. Masaeli, E.; et al. Fabrication, characterization and cellular compatibility of poly(hydroxy alkanoate) composite nanofibrous scaffolds for nerve tissue engineering. *Plos One.* **2013,** *8(2),* e57157.

107. Zhao, K.; Deng, Y.; Chun Chen, J.; Chen, G. Q.; Polyhydroxyalkanoate (PHA) scaffolds with good mechanical properties and biocompatibility. *Biomater.* **2003,** *24(6),* 1041–1045.

108. Cheng, S. T.; Chen, Z. F.; and Chen, G. Q.; The expression of cross-linked elastin by rabbit blood vessel smooth muscle cells cultured in polyhydroxyalkanoate scaffolds. *Biomater.* **2008,** *29(31),* 4187–4194.

109. Francis, L.; Meng, D.; Knowles, J. C.; Roy, I.; and Boccaccini, A. R.; Multi-functional P(3HB) microsphere/45S5 Bioglass-based composite scaffolds for bone tissue engineering. *Acta Biomater.* **2010,** *6(7),* 2773–2786.

110. Misra, S. K.; et al. Poly(3-hydroxybutyrate) multifunctional composite scaffolds for tissue engineering applications. *Biomater.* **2010,** *31(10),* 2806–2815.

111. Lootz, D.; Behrend, D.; Kramer, S.; Freier, T.; Haubold, A.; Benkiesser, G.; Schmitz, K. P.; Becher, B.; Laser cutting: influence on morphological and physicochemical properties of polyhydroxybutyrate. *Biomater.* **2001,** *22(18),* 2447–2452.

112. Fischer, D.; Li, Y.; Ahlemeyer, B.; Kriglstein, J.; and Kissel, T.; In vitro cytotoxicity testing of polycations: influence of polymer structure on cell viability and hemolysis. *Biomater.* **2003,** *24(7),* 1121–1131.

113. Nitschke, M.; Schmack, G.; Janke, A.; Simon, F.; Pleul, D.; and Werner, C.; Low pressure plasma treatment of poly(3-hydroxybutyrate): toward tailored polymer surfaces for tissue engineering scaffolds. *J. Biomed. Mater. Res.* **2002,** *59(4),* 632–638.

114. Chanvel-Lesrat, D. J.; Pellen-Mussi, P.; Auroy, P.; and Bonnaure-Mallet, M.; Evaluation of the in vitro biocompatibility of various elastomers. *Biomater.* **1999,** *20,* 291–299.

115. Boyan, B. D.; Hummert, T. W.; Dean, D. D.; and Schwartz, Z.; Role of material surfaces in regulating bone and cartilage cell response. *Biomater.* **1996,** *17,* 137–146.

116. Bowers, K. T.; Keller, J. C.; Randolph, B. A.; Wick, D. G.; and Michaels, C. M.; Optimization of surface micromorphology for enhanced osteoblasts responses in vitro. *Int. J. Oral. Max. Impl.* **1992,** *7,* 302–310.

117. Cochran, D.; Simpson, J.; Weber, H.; and Buser, D.; Attachment and growth of periodontal cells on smooth and rough titanium. *Int. J. Oral. Max. Impl.* **1994,** *9,* 289–297.

118. Bagrov, D. V.; et al. Amorphous and semicrystalline phases in ultrathin films of poly(3-hydroxybutirate) *Tech. Proc. 2012 NSTI Nanotechnol. Conf. Expo, NSTI-Nanotech.* **2012,** 602–605.

119. Sevastianov, V. I.; Perova, N. V.; Shishatskaya, E. I.; Kalacheva, G. S.; and Volova, T. G.; Production of purified polyhydroxyalkanoates (PHAs) for applications in contact with blood. *J. Biomater. Sci. Polym. Ed.* **2003,** *14,* 1029–1042.

120. Seebach, D.; Brunner, A.; Burger, H. M.; Schneider, J.; and Reusch, R. N.; Isolation and 1H-NMR spectroscopic identification of poly(3-hydroxybutanoate) from prokaryotic and eukaryotic organisms. Determination of the absolute configuration (R) of the monomeric unit 3-hydroxybutanoic acid from Escherichia coli and spinach. *Eur. J. Biochem.* **1994,** *224(2),* 317–328.

121. Myshkina, V. L.; Nikolaeva, D. A.; Makhina, T. K.; Bonartsev, A. P.; and Bonartseva, G. A.; Effect of growth conditions on the molecular weight of poly-3-hydroxybutyrate produced by Azotobacter chroococcum 7B. *Appl. Biochem. Microbiol.* **2008,** *44(5),* 482–486.

122. Myshkina, V. L.; et al. Biosynthesis of poly-3-hydroxybutyrate-3-hydroxyvalerate copolymer by Azotobacter chroococcum strain 7B. *Appl. Biochem. Microbiol.* **2010,** *46(3),* 289–296.

123. Reusch, R. N.; Poly-β-hydroxybutryate/calcium polyphosphate complexes in eukaryotic membranes. *Proc. Soc. Exp. Biol. Med.* **1989,** *191,* 377–381.

124. Reusch, R. N.; Biological complexes of poly-β-hydroxybutyrate. *FEMS Microbiol. Rev.* **1992,** *103,* 119–130.

125. Reusch, R. N.; Low molecular weight complexed poly(3-hydroxybutyrate): a dynamic and versatile molecule in vivo. *Can. J. Microbiol.* **1995,** *41*(Suppl. 1), 50–54.

126. Müller, H. M.; and Seebach, D.; Polyhydroxyalkanoates: a fifth class of physiologically important organic biopolymers? *Angew Chemie.* **1994,** *32,* 477–502.

127. Huang, R.; and Reusch, R. N.; Poly(3-hydroxybutyrate) is associated with specific proteins in the cytoplasm and membranes of Escherichia coli. *J. Biol. Chem.* **1996,** *271,* 22196–22201.

128. Reusch, R. N.; Bryant, E. M.; and Henry, D. N.; Increased poly-(R)-3-hydroxybutyrate concentrations in streptozotocin (STZ) diabetic rats. *Acta Diabetol.* **2003,** *40(2),* 91–94.

129. Reusch, R. N.; Sparrow, A. W.; and Gardiner, J.; Transport of poly-β-hydroxybutyrate in human plasma. *Biochim. Biophys. Acta.* **1992,** *1123,* 33–40.

130. Reusch, R. N.; Huang, R.; and Kosk-Kosicka, D.; Novel components and enzymatic activities of the human erythrocyte plasma membrane calcium pump. *FEBS Lett.* **1997,** *412(3),* 592–596.

131. Pavlov, E.; et al. Large, voltage-dependent channel, isolated from mitochondria by water-free chloroform extraction. *Biophys. J.,* **2005,** *88,* 2614–2625.

132. Theodorou, M. C.; Panagiotidis, C. A.; Panagiotidis, C. H.; Pantazaki, A. A.; and Kyriakidis, D. A.; Involvement of the AtoS-AtoC signal transduction system in poly-(R)-3-hydroxybutyrate biosynthesis in Escherichia coli. *Biochim. Biophys. Acta.* **2006,** *1760(6),* 896–906.

133. Wiggam, M. I.; et al. Treatment of diabetic ketoacidosis using normalization of blood 3-hydroxy-butyrate concentration as the endpoint of emergency management. *Diabetes Care.* **1997,** *20,* 1347–1352.

134. Larsen, T.; and Nielsen, N. I.; Fluorometric determination of beta-hydroxybutyrate in milk and blood plasma. *J. Dairy Sci.* **2005,** *88(6),* 2004–2009.

135. Agrawal, C. M.; Athanasiou, K. A.; Technique to control pH in vicinity of biodegrading PLA-PGA implants. *J. Biomed. Mater. Res.* **1997,** *38(2),* 105–114.

136. Ignatius, A. A., Claes, L. E.; In vitro biocompatibility of bioresorbable polymers: poly(l, dl-lactide) and poly(l-lactide-co-glycolide). **1996,** *17(8),* 831–839.

137. Rihova, B.; Biocompatibility of biomaterials: hemocompatibility, immunocompatibility and biocompatibility of solid polymeric materials and soluble targetable polymeric carriers. *Adv. Drug. Delivery Rev.* **1996,** *21,* 157–176.

138. Ceonzo, K.; Gaynor, A.; Shaffer, L.; Kojima, K.; Vacanti, C. A.; and Stahl, G. L.; Polyglycolic acid-induced inflammation: role of hydrolysis and resulting complement activation. *Tissue Eng.* **2006,** *12(2),* 301–308.

139. Chasin, M.; and Langer, R.; Eds. Biodegradable Polymers as Drug Delivery Systems, New York: Marcel Dekker; **1990.**

140. Johnson, O. L.; and Tracy, M. A.; Peptide and protein drug delivery. In: Ed. Mathiowitz, E.; Encyclopedia of Controlled Drug Delivery. Hoboken, New Jersey: John Wiley and Sons; **1999,** *2,* 816–832.

141. Jain, R. A.; The manufacturing techniques of various drug loaded biodegradable poly(lactide-co-glycolide) (PLGA) devices. *Biomater.* **2000,** *21,* 2475–2490.

142. Gursel, I.; and Hasirci, V.; Properties and drug release behaviour of poly(3-hydroxybutyric acid) and various poly(3-hydroxybutyrate-hydroxyvalerate) copolymer microcapsules. *J. Microencapsul.,* **1995,** *12(2),* 185–193.

143. Li, J.; Li, X.; Ni, X.; Wang, X.; Li H.; and Leong, K. W.; Self-assembled supramolecular hydrogels formed by biodegradable PEO–PHB–PEO triblock copolymers and a-cyclodextrin for controlled drug delivery. *Biomater.* **2006,** *27,* 4132–4140.

144. Akhtar, S.; Pouton, C. W.; Notarianni, L. J.; Crystallization behaviour and drug release from bacterial polyhydroxyalkanoates. *Polym.* **1992,** *33(1),* 117–126.

145. Akhtar, S.; Pouton, C. W.; and Notarianni, L. J.; The influence of crystalline morphology and copolymer composition on drug release from solution cast and melting processed P(HB-HV) copolymer matrices. *J. Controlled Release.* **1991,** *17,* 225–234.

146. Korsatko, W.; Wabnegg, B.; Tillian, H. M.; Braunegg, G.; and Lafferty, R. M.; Poly-D-hydroxybutyric acid-a biologically degradable vehicle to regard release of a drug. *Pharm. Ind.* **1983,** *45,* 1004–1007.

147. Korsatko, W.; Wabnegg, B.; Tillian, H. M.; Egger, G.; Pfragner, R.; and Walser, V.; Poly D(-)-3-hydroxybutyric acid (poly-HBA)-a biodegradable former for long-term medication dosage. 3. Studies on compatibility of poly-HBA implantation tablets in tissue culture and animals. *Pharm. Ind.* **1984,** *46,* 952–954.

148. Kassab, A. C.; Piskin, E.; Bilgic, S.; Denkbas, E. B.; and Xu, K.; Embolization with polyhydroxybutyrate (PHB) micromerspheres: in vivo studies, *J. Bioact. Compat. Polym.* **1999,** *14,* 291–303.

149. Kassab, A. C.; Xu, K.; Denkbas, E. B.; Dou, Y.; Zhao, S.; and Piskin, E.; Rifampicin carrying polyhydroxybutyrate microspheres as a potential chemoembolization agent. *J. Biomater. Sci. Polym. Ed.* **1997,** *8,* 947–961.

150. Sendil, D.; Gursel, I.; Wise, D. L.; and Hasirci, V.; *Antibiotic release from biodegradable PHBV microparticles. J. Control. Release.* **1999,** *59,* 207–17.

151. Gursel, I.; Yagmurlu, F.; Korkusuz, F.; and Hasirci, V.; In vitro antibiotic release from poly(3-hydroxybutyrate-co-3-hydroxyvalerate) rods. *J. Microencapsul.* **2002,** *19,* 153–164.

152. Turesin, F.; Gursel, I.; and Hasirci, V.; Biodegradable polyhydroxyalkanoate implants for osteomyelitis therapy: in vitro antibiotic release. *J. Biomater. Sci. Polym. Ed.* **2001,** *12,* 195–207.

153. Turesin, F.; Gumusyazici, Z.; Kok, F. M.; Gursel, I.; Alaeddinoglu, N. G.; and Hasirci, V.; Biosynthesis of polyhydroxybutyrate and its copolymers and their use in controlled drug release. *Turk. J. Med. Sci.* **2000,** *30,* 535–541.

154. Gursel, I.; Korkusuz, F.; Turesin, F.; Alaeddinoglu, N. G.; and Hasirci, V.; In vivo application of biodegradable controlled antibiotic release systems for the treatment of implant-related osteomyelitis. *Biomaterials,* **2001,** *22(1),* 73–80.

155. Korkusuz, F.; Korkusuz, P.; Eksioglu, F.; Gursel, I.; and Hasirci, V.; In vivo response to biodegradable controlled antibiotic release systems. *J. Biomed. Mater. Res.* **2001,** *55(2),* 217–228.

156. Yagmurlu, M. F.; Korkusuz, F.; Gursel, I.; Korkusuz, P.; Ors. U.; and Hasirci, V.; Sulbactam-cefoperazone polyhydroxybutyrate-co-hydroxyvalerate (PHBV) local antibiotic delivery system: In vivo effectiveness and biocompatibility in the treatment of implantrelated experimental osteomyelitis. *J. Biomed. Mater. Res.* **1999,** *46*: 494–503.

157. Yang, C.; Plackett, D.; Needham, D.; and Burt, H. M.; PLGA and PHBV microsphere formulations and solid-state characterization: possible implications for local delivery of fusidic acid for the treatment and prevention of orthopaedic infections. *Pharm. Res.* **2009,** *26(7),* 1644–1656.

158. Kosenko, R. Yu.; Iordanskii, A. L.; Markin, V. S.; Arthanarivaran, G.; Bonartsev, A. P.; and Bonartseva, G. A.; Controlled release of antiseptic drug from poly(3-hydroxybutyrate)-based membranes. combination of diffusion and kinetic mechanisms. *Pharmaceutical Chem. J.* **2007,** *41(12),* 652–655.

159. Khang, G.; Kim, S. W.; Cho, J. C.; Rhee, J. M.; Yoon, S. C.; and Lee, H. B.; Preparation and characterization of poly(3-hydroxybutyrate-co-3-hydroxyvalerate) microspheres for the sustained release of 5-fluorouracil. *Biomed. Mater. Eng.,* **2001,** *11:* 89–103.

160. Bonartsev, A. P.; et al. Sustained release of the antitumor drug paclitaxel from poly(3-hydroxybutyrate)-based microspheres. *Biochem. (Moscow) Suppl. Ser. B: Biomed.l Chem.* **2012,** *6(1),* 42–47.

161. Yakovlev, S. G.; Bonartsev, A. P.; Boskhomdzhiev, A. P.; Bagrov, D. V.; Efremov, Yu. M.; Filatova, E. V.; Ivanov, P. V.; Mahina, T. K.; Bonartseva, G. A.; In vitro cytotoxic activity of poly(3-hydroxybutyrate) nanoparticles loaded with antitumor drug paclitaxel. Technical Proceedings of the 2012 NSTI Nanotechnology Conference and Expo, NSTI-Nanotech **2012,** 190–193.

162. Shishatskaya, E. I.; Goreva, A. V.; Voinova, O. N.; Inzhevatkin, E. V.; Khlebopros, R. G.; Volova, T. G.; Evaluation of antitumor activity of rubomycin deposited in absorbable polymeric microparticles. *Bull. Exp. Biol. Med.* **2008,** *145(3),* 358–361.

163. Filatova, E. V.; Yakovlev, S. G.; Bonartsev, A. P.; Mahina, T. K.; Myshkina, V. L.; and Bonartseva, G. A.; Prolonged release of chlorambucil and etoposide from poly-3-oxybutyrate-based microspheres *Appl. Biochem. Microbiol.* **2012,** *48(6),* 598–602.

164. Bonartsev, A. P.; et al. New poly-(3-hydroxybutyrate)-based systems for controlled release of dipyridamole and indomethacin. *Appl. Biochem. Microbiol.* **2006,** *42(6),* 625–630.

165. Coimbra, P. A.; De Sousa, H. C.; Gil, M. H.; Preparation and characterization of flurbiprofen-loaded poly(3-hydroxybutyrate-co-3-hydroxyvalerate) microspheres. *J. Microencapsul.* **2008,** *25(3),* 170–178.

166. Wang, C.; Ye, W.; Zheng, Y.; Liu, X.; Tong, Z.; Fabrication of drug-loaded biodegradable microcapsules for controlled release by combination of solvent evaporation and layer-by-layer self-assembly. Int. *J. Pharm.* **2007,** *338(1–2),* 165–173.

167. Salman, M. A.; Sahin, A.; Onur, M. A.; Oge, K.; Kassab, A.; Aypar, U.; Tramadol encapsulated into polyhydroxybutyrate microspheres: in vitro release and epidural analgesic effect in rats. *Acta Anaesthesiol. Scand.* **2003,** *47,* 1006–1012.

168. Bonartsev, A. P.; Livshits, V. A.; Makhina, T. A.; Myshkina, V. L.; Bonartseva, G. A.; and Iordanskii, A. L.; Controlled release profiles of dipyridamole from biodegradable microspheres on the base of poly(3-hydroxybutyrate). *Express Polym. Lett.* **2007,** *1(12),* 797–803.

169. Livshits, V. A.; et al. Microspheres based on poly(3-hydroxy)butyrate for prolonged drug release. *Polym. Sci. Series B.* **2009,** *51(7–8),* 256–263.

170. Bonartsev, A. P.; Postnikov, A. B., Myshkina, V. L.; Artemieva, M. M.; and Medvedeva, N. A.; A new system of nitric oxide donor prolonged delivery on basis of controlled-release polymer, polyhydroxybutyrate. *Am. J. Hypert.* **2005,** *18(5A),* p.A

171. Bonartsev, A. P.; et al. A new in vivo model of prolonged local nitric oxide action on arteries on basis of biocompatible polymer. *The J. Clinical Hypertension.* **2007,** Suppl. A, *9(5),* A152(c).

172. Stefanescu, E. A.; Stefanescu, C.; Negulescu, I. I.; Biodegradable polymeric capsules obtained via room temperature spray drying: preparation and characterization. *J. Biomater. Appl.,* **2011,** *25(8),* 825–849.

173. Costa, M. S.; Duarte, A. R.; Cardoso, M. M.; and Duarte, C. M.; Supercritical antisolvent precipitation of PHBV microparticles. *Int J. Pharm.* **2007**, *328(1)*, 72–77.

174. Riekes, M. K.; Junior, L. R.; Pereira, R. N.; Borba, P. A.; Fernandes, D.; and Stulzer, H. K.; Development and evaluation of poly(3-hydroxybutyrate-co-3-hydroxyvalerate) and polycaprolactone microparticles of nimodipine. *Curr Pharm Des.* **2013** March **12**. [Epub ahead of print].

175. Bazzo, G. C.; et al. Enhancement of felodipine dissolution rate through its incorporation into Eudragit® E-PHB polymeric microparticles: in vitro characterization and investigation of absorption in rats. *J. Pharm. Sci.* **2012**, *101(4)*, 1518–1523.

176. Zhu, X. H.; Wang, C. H.; Tong, Y. W.; In vitro characterization of hepatocyte growth factor release from PHBV/PLGA microsphere scaffold. *J. Biomed. Mater. Res A.,* **2009**, *89(2)*, 411–423.

177. Parlane, N. A.; et al. Vaccines displaying mycobacterial proteins on biopolyester beads stimulate cellular immunity and induce protection against tuberculosis. *Clin. Vaccine Immunol.* **2012**, *19(1)*, 37–44.

178. Yilgor, P.; Tuzlakoglu, K.; Reis, R. L.; Hasirci, N.; and Hasirci, V.; Incorporation of a sequential BMP-2/BMP-7 delivery system into chitosan-based scaffolds for bone tissue engineering. *Biomaterials.* **2009**, *30(21)*, 3551–3559.

179. Errico, C.; Bartoli, C.; Chiellini, F.; and Chiellini, E.; Poly(hydroxyalkanoates)-based polymeric nanoparticles for drug delivery. *J. Biomed. Biotechnol.* **2009**, 571702.

180. Althuri, A.; Mathew, J.; Sindhu, R.; Banerjee, R.; Pandey, A.; Binod, P.; Microbial synthesis of poly-3-hydroxybutyrate and its application as targeted drug delivery vehicle. *Bioresour Technol.* **2013**, pii: S0960-8524(13)00129-6.

181. Pouton, C. W.; and Akhtar, S.; Biosynthetic polyhydroxyalkanoates and their potential in drug delivery. *Adv. Drug Deliver. Rev.* **1996**, *18,* 133–162.

CHAPTER 2

TRENDS IN NEW GENERATION OF BIODEGRADABLE POLYMERS (PART 2)

A. L. IORDANSKII, S. V. FOMIN, A. A. BURKOV, YU. N. PANKOVA, and G. E. ZAIKOV

CONTENTS

2.1   INTRODUCTION

One of the most pressing problems facing humanity is increasing pollution. In this regard, question of recycling of synthetic polymeric materials arises very sharply. Plastics production rates are growing exponentially and the production volumes 100 millions of tons annually. One of the most dynamic areas of the use of plastics is packaging industry (40–50% of the total production of plastics). Thus, billions of tons of municipal solid waste products constitute more than half of the short-term or one-time application on the basis of large common polyolefins. A possible solution of this problem is to create a biodegradable polymer composition. The fastest growing trend in this area is the use of polyhydroxyalkanoates (PHAs). Some physical and chemical characteristics of PHA are similar to these of synthetic polymers (polypropylene, polyethylene).

However, in addition to their thermoplasticity, representatives of PHAs have optical activity, increase induction period of oxidation, exhibit the piezoelectric effect and, what is most important, they are characterized as being biodegradable and biocompatible. At the same time, the PHAs have disadvantages (high cost, brittleness), which can be partially or completely compensated by using composite materials based on blends with other polymers, with dispersed fillers or plasticizers. Taking into account all the above, we have suggested to create a mixed polymer composite based on poly-3-hydroxybutyrate (PHB) and polyisobutylene (PIB).

2.2   OBJECTS AND METHODS

The objects of the study were high molecular weight PIB of mark "P-200" and PHB Lot 16F. The poly-3-hydroxybutyrate was obtained by microbiological synthesis in company "BIOMER®" (Germany). PHB is a white fine powder; the density is 1.25 g/cm³, and the molecular weight is 325 kDa. Polyisobutylene of high molecular weight "P-200" is a white elastic material, transparent in thin films, odorless, with density of 0.93 g/cm³, and with molecular weight of about 175–225 kDa.

These materials have been chosen due to their economic expediency and the valuable combination of physical and chemical, physical and mechanical, and other properties of individual polymers-PHB is a brittle thermoplastic; PIB an elastomer. Preparation of composite materials based on combinations of plastics and elastomers is well known: plastics are used as polymeric fillers in elastomers, improving their technological and working characteristics; elastomers effectively improve strike viscosity and reduce brittleness in compositions based on plastics.

The polymer compositions were prepared in the following proportions: PHB:PIB=10:90, 20:80, 30:70, 40:60, 50:50, and 60:40 (by weight ratio here and hereinafter). Then, the individual polymers were investigated. The first stage of blending was carried out on laboratory mixing rolls with the roll length of 320 mm and diameter of 160 mm. Temperature of the back roll was 60°C, temperature of the front roll was 50°C. In these conditions, the compositions were obtained, in which the micropowder of PHB was distributed in the continuous matrix of PIB. Deeper joining, mutual segmental solubility of polymers require an additional high temperature treatment of compositions, so the second stage of the composition processing was carried out in plunger extruder at the temperature of 185°C.

Average molecular weight of the elastomer was characterized by viscometry method. Average molecular weight of PIB was calculated according to the equation Mark-Houwink:

$$[\eta] = \kappa \times Mm^{\alpha}, \tag{2.1}$$

[η]—the intrinsic viscosity, ml/g;
Mm—average molecular weight of the polymer;
K, α—constants for a given system "polymer-solvent".

Heptane was used as solvent, the constants—K=$1.58*10^{-4}$, α=0.69 [8]. For each sample, a series of solutions were prepared with different concentrations (0.2, 0.4, 0.6, 0.8, and 1.0 g/100 ml), and then determines the relative, specific, and intrinsic viscosity. The calculated molecular weights were presented in Table 2.1.

Investigation of the structure of the compositions by atomic force microscopy was performed on the tunnel atomic force microscope brands Ntegra Prima (company "NT–MDT") (cantilever NSG-01 with a frequency of 0.5–1 Hz). Investigations were carried out by semicontact mode. Microtome cuts were made on the brand Microm HM-525 (company "Thermoscientific," Germany).

Investigation of the structure of the compositions was carried out by differential scanning calorimetry on differential scanning calorimeter DSC-60 (company "Shimadzu," Japan). Samples of PHB–PIB blends weighing several milligrams were placed in open aluminium crucibles with a diameter of 5.8 mm, a height of 1.5 mm, and weighing 13 mg upper temperature limit of 600°C. Temperature ranges from −100 to +250°C, heating rate of 10°/min. Liquid nitrogen is used to generate low temperature. Instrument calibration was performed according to indium, tin, and lead.

The rheological curves of pure polymer melts and their mixtures were obtained with the multifunction rheometer "StressTech" (company "REOLOGICA Instruments AB," Sweden). The measuring cell consisted of two parallel planes (the lower plane was fixed, the upper plane being a rotating rotor); the shear rate was 0.1 s⁻¹, the temperature range was 443–513 K (170–240°C). The lower limit of the temperature range was chosen by the melting temperature of PHB (174.4°C—the data obtained for DSC), the upper limit is the beginning of irreversible degradation in polymers.

Polymer samples were subjected to soil degradation in a laboratory at a temperature 22–25°C. Samples in the form of a film thickness of 50 μm was placed in the soil to a depth of 1.5–2 cm. Biodegradation rate was assessed by evaluating the mass loss of the samples. Mass loss was fixed by weighing the samples on an analytical balance.

2.3   RESULTS AND DISCUSSION

Properties of mixed polymer compositions are determined by many factors, among which in the first place should be allocated phase structure (ratio and the size of the phase domains). Therefore, at the first stage of the research attention has been paid to study the structure of formed compositions. In the investigation of samples with a low content of PHB (10–30% by weight) has been found that it forms a discontinuous phase, i.e., distributed in a continuous matrix PIB as separate inclusions of the order of 1–2 μm. The results of atomic force microscopy for the composition ratio of PHB–PIB 20:80 were shown in Figure 2.1.

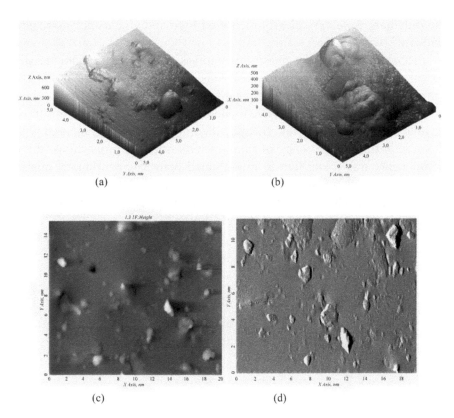

(a)   (b)

(c)   (d)

FIGURE 2.1   Photomicrographs of the relief of films with PHB–PIB ratio of 20:80 (scan size 5 × 5 um) (a, b) and AFM phase contrast image of the film with a ratio of 20:80 PHB–PIB (scan size 20 × 20 um) (c, d).

Atomic force microscopy is considered one of the most perspective methods for studying polymer blends is because this method allows to clearly define the phase boundary and its scale. Polymer identification was performed by controlling the interaction of the probe with the sample surface at different points. When approaching the surface of the cantilever is deflected downward (to the sample) due to attractive forces until the probe comes into contact with the sample. When the probe is withdrawn from the studied surface, a hysteresis is observed, associated with the adhesive forces. Adhesion forces between the probe and the sample are forcing them to remain in contact, which causes the cantilever to bend. Phase of the polymers were very clearly identified by mapping curves "approach-removal" of the probe.

Probable range of the phase inversion is an important characteristic for mixed biodegradable composites. This allows from a practical point of view to establish the minimum concentration of PHB (at which a continuous phase is formed) for intensive biodegradation as microorganisms must be able to penetrate deep into the mixed composite. When compared with the atomic force microscopy of samples with different proportions of the components was found that in investigated materials continuous matrix formation occurs when PHB content in the mixture of about 40–50 per cent by weight. The results of atomic force microscopy for the composition ratio of PHB–PIB 50:50 shown in Figure 2.2.

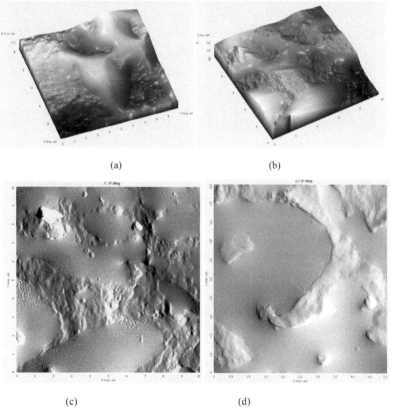

(a)   (b)

(c)   (d)

FIGURE 2.2 Photomicrographs of the relief of films with PHB–PIB ratio of 50:50 (scan size 10 × 10 μm) (a, b) and AFM phase contrast image of the film with a ratio of 50:50 PHB–PIB ((scan size 10 × 10 μm) (c) and 5 × 5 μm) (d).

Thus, by atomic force microscopy were determined scale structures and distributions of polymers according to the ratio of the starting components in the mixture, and the approximate range of probable phase inversion (about 40–50% by weight of PHB).

Determination of the glass transition temperature is an informative method of research, the phase structure of polymer blends. In the case of mixed compositions, glass transition may occur in each phase separately, if polymers do not interact with each other. In another case glass transition in mixtures is fully cooperative process, involving macromolecules mixed polymer segments. This composition would have single glass transition temperature, which varies monotonically depending on the mixture composition [1]. The glass transition temperatures of polymer components of the mixtures were determined by differential scanning calorimetry. The results of DSC shown in Figure 2.3.

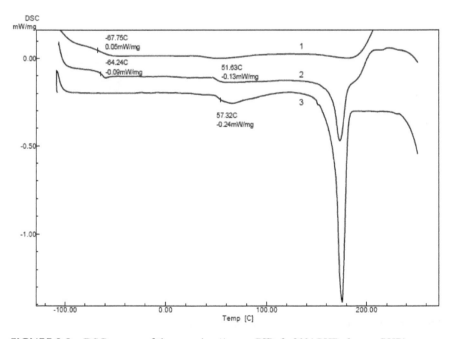

FIGURE 2.3   DSC curves of the samples (1-pure PIB, 2–20% PHB, 3-pure PHB).

In all cases, the specific heat jump was recorded, which corresponds to the glass transition of phases polymers. It was noted that displacement values of glass transition temperature of phase PIB to higher temperatures with increasing content of PHB (from −68°C for pure PIB to −64°C for the composition with 60 per cent by weight of the PHB). For PHB phase displacement of values of glass transition temperature is shown in Figure 2.4.

As can be seen from the data presented, the displacement of values of glass transition temperature of PHB occurs about 6–7°C. Most often in the literature as reasons for this phenomenon is called a limited solubility of the mixture components in each other (from a fraction of a few per cent) [1–3]. However, for the polymers probability such variant is extremely small because of the significant thermodynamic incompatibility. More probable reasons of displacement of the glass transition temperature may be changes of the supramolecular structure of polymers when mixing; as well as differences in thermal expansion coefficients of polymers in the region above and below the glass transition temperature.

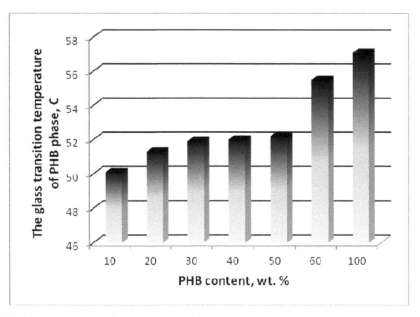

FIGURE 2.4   Changes in the glass transition temperature of phase PHB.

The glass transition temperature change deserves attention because is a sharp jump in the of PHB content of 50–60 per cent by weight. This phenomenon may also be due to the probable phase inversion (previously established by microscopy methods) at said ratio of the components. Continuous structure more rigid of PHB has a higher glass transition temperature than the individual inclusions of thermoplastic material, isolated from each other by a continuous matrix of PIB. Thus, as a result of studies on the structure of compositions of PHB–PIB was confirmed formation of a heterogeneous two-phase system. For biodegradable polymer compositions, it may be advantage, because this system is more susceptible to external influences destructive.

However, for practical use of PHB–PIB compositions is necessary to evaluate the possibility of processing these materials into finished products. As most polymer blends are processed by melting them, investigations of rheological properties of these compositions are of great scientific and practical interest. Information about the structure of composition that can be obtained on the basis of rheological investigation is the level of intermolecular interactions, the degree of macromolecules ordering, the phase structure of polymer blends. Dependence "melt viscosity—temperature" were investigated for all compositions. The viscosity values at various temperatures for blends of polymers with different ratios are shown in Figure 2.5.

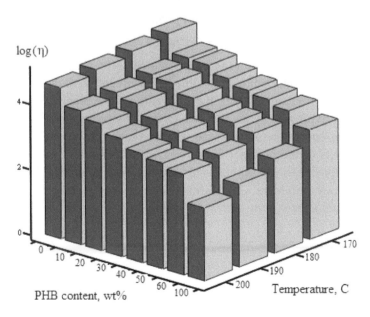

FIGURE 2.5   The viscosity values at various temperatures for blends of polymers with different ratios.

The viscosity values of the composition decreases considerably with increasing temperature and increasing the content of PHB that gives the chance regulation of technological properties of compositions in a wide range. Based on the significant difference of the solubility parameters calculated previously [5-6], it seems reasonable assumption on the formation of a biphasic system melts of mixtures of these polymers. The assumption of formation of two-phase mixtures of melts of these polymers seems justified based on significant difference in the solubility parameters, which calculated previously [6]. Kerner–Takayanagi equation applicable to describe the rheological properties of two-phase mixtures of polymers [1]:

$$\eta = \eta_m \times \frac{(7 - 5v_m) \times \eta_m + (8 - 10v_m) \times \eta_f - (7 - 5v_m) \times (\eta_m - \eta_f) \times f_f}{(7 - 5v_m) \times \eta_m + (8 - 10v_m) \times \eta_f - (8 - 10v_m) \times (\eta_m - \eta_f) \times f_f}, \qquad (2.2)$$

η—viscosity of the mixture, Pa×s;

η_m—viscosity of the matrix, Pa×s;

η_f—viscosity of the dispersed phase, Pa×s;

φ_f—volume fraction of the dispersed phase;

v_m—Poisson's ratio of the matrix (assumed equal to 0.5).

This expression describes the dependence of the properties of the mixture composition, excluding inevitable phase inversion in the mixture. According to Ref. (Paul, 2009a), the viscosity of the blend composition is graphically expressed by the two curves corresponding to the two limiting cases when one or the other phase is continuous throughout the range of compositions of the polymer mixture. Theoretical values of the viscosity of the compositions of various compositions were calculated based on the viscosity base polymers. Calculations on model of Kerner-Takayanagi at different temperatures are shown in Figure 2.6.

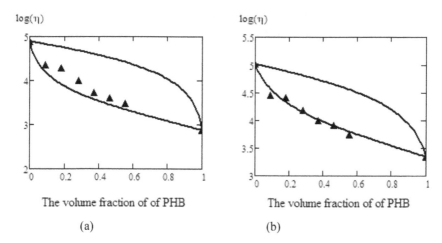

FIGURE 2.6   Viscosity of the composition at a temperature of 180°C (a) and 170°C (b) (theoretical calculations on model of Kerner–Takayanagi; ▲ experimental data).

By analyzing data of Figure 2.6, it was found that PHB forms a continuous matrix in the molten polymer at any ratio. The experimental values of the viscosity are close to the bottom theoretical curve, which corresponds to the calculation for the case of the formation of the matrix of PHB. Thus, during the melting of PHB observed phase inversion phenomenon in accordance with the laws of Ref. [4, 6] more fluid melt PHB forms a continuous phase in the entire range of concentrations.

Basic rheological parameters of polymer blends determined by the properties of the polymer matrix, so the properties of melts are generally defined by PHB. Lower viscosity of the melt blend composition (due to the formation of a continuous matrix PHB) in this case will significantly simplify the processing of the investigated materials. Also, the data in Figure 2.6 are consistent with the position that the greater the difference in viscosity of mixed polymers, the earlier the formation of a continuous matrix less viscous component.

Previous studies [7] revealed that the main physical and mechanical characteristics (tensile strength, tensile modulus, and elongation at break), PHB–PIB composition containing 30–50 per cent by weight of PHB, not inferior to traditional polymers used in the production of packaging (polyethylenes of different brands). But the undeniable advantage of the compositions PHB–PIB is the possibility of biodegradation.

It has been found that mixtures containing 40–60 per cent PHB stepwise lose mass when exposed to soil. In the first stage (100 days), intensive degradation of phase PHB occurs under the action of microorganisms. The result of the first stage of this scenario is PIB samples with very high surface area. In the second step (after 100 days), mass loss is stabilized and is much slower.

TABLE 2.1   The average molecular weight of PIB samples

PIB samples	The intrinsic viscosity	The average molecular weight
The original sample	0.68	185 000
The sample after 200 days of Exposure in the soil (0% PHB)	0.65	175 000
The sample after 200 days of Exposure in the soil (60% PHB)	0.49	115 000

As can be seen from Table 2.1, the average molecular weight PIB sample (previously containing PHB) much less compared to the original molecular weight PIB and PIB after exposure to soil. Significant impact of PHB phase on the ability of PIB to degradation in soil was found by research of the molecular weight of the elastomer. Destruction of the original structure, PIB can occur due to physical factors (large specific surface films), chemical factors (interaction of the products of biodegradation of PHB with PIB phase, and initiation of the degradation of the elastomer) and biological factors (accumulation of microorganisms biomass on the surface of mixed films and therefore, some microorganisms can use the matrix PIB as a carbonaceous substrate for growth.

2.4   CONCLUSION

There are some conclusions drawn in the course of studying structure and properties of polymer system "PHB–PIB":
1.   Structure of mixed compositions of PHB–PIB investigated by atomic force microscopy; confirmed the formation of a heterogeneous two-phase structure. PHB forms a continuous elongated

structure in the matrix of PIB when content of PHB in the mixture is greater than 30 per cent.

2. Displacement of glass transition temperatures of individual phases in a mixture of polymers determined by differential scanning calorimetry. This may be indicative of their limited interaction, despite the significant difference of the solubility parameters.

3. Comparison of experimental and theoretical values of viscosity confirmed the formation of a continuous matrix less viscous PHB. This greatly simplifies the processing of melts of compositions PHB–PIB.

4. Microbial degradation of PHB phase occurs almost entirely within 100–125 days. The remaining friable matrix PIB are much more susceptible to degradation by other destructive factors (oxygen of the air, temperature, and mechanical stress).

5. Process of biodegradation of PHB phase influences on the structure of PIB. This is reflected in a sharp decrease of average molecular weight of PIB.

ACKNOWLEDGEMENTS

The work was supported by the Russian Foundation for Basic Research (grant no. 13-03-00405-a) and the Russian Academy of Sciences under the program "Construction of New Generation Macromolecular Structures" (03/OC-14).

KEYWORDS

- AFM
- Blends
- Degradation
- DSC
- Morphology
- Poly(3-hydroxybutyrate)
- Polyisobutylene
- Rheology
- Soil
- Weight loss

REFERENCES

1. Kuleznev, V. N.; Mixtures of the Polymers. Moscow; **1980**.
2. Paul, D.; and Bucknell, K.; Polyblend Compounds. Functional Properties. Saint Petersburg; **2009**.
3. Paul, D.; and Bucknell, K.; Polyblend Compounds. Saint Petersburg: Systematics; **2009**.
4. Malkin, A.; Rheology: Concepts, Methods, and Applications. St. Petersburg; **2007**.
5. Iordanskii, A. L.; Fomin, S. V.; and Burkov, A. A.; Structure and melt rheology mixed compositions of polyisobutylene and poly-3-hydroxybutyrate. *Plastic Mater.* **2012**, *7(7)*, 13–16.
6. Schramm, G.; Basics Practical Rheology and Rheometry. Moscow; **2003**.
7. Iordanskii, A. L.; Fomin, S. V.; and Burkov, A. A.; Investigation of the structure and properties of biodegradable polymer composites based on poly-3-hydroxybutyrate and polyisobutylene. Bull. Kazan University Technol. **2013**, *16(9)*, 115–119.
8. Sangalov, Y. A.; Polymers and Copolymers of Isobutylene: Fundamental and Applied Aspects of the Problem. Ufa; **2001**.

CHAPTER 3

TRENDS IN NEW GENERATION OF BIODEGRADABLE POLYMERS (PART 3)

A. P. BONARTSEV, A. P. BOSKHOMODGIEV, A. L. IORDANSKII,
G. A. BONARTSEVA, A. V. REBROV T.K. MAKHINA,
V.L. MYSHKINA, S. A. YAKOVLEV, E. A. FILATOVA,
E. A. IVANOV, D. V. BAGROV, G. E. ZAIKOV, and M. I. ARTSIS

CONTENTS

3.1 INTRODUCTION

This work is designed to be an informative source for biodegradable poly(3-hydroxybutyrate) and its derivatives' research. We focus on hydrolytic degradation kinetics at 37 and 70°C in phosphate buffer to compare PLA and PHB kinetic profiles. Besides, we reveal the kinetic behavior for copolymer PHBV (20% of 3-hydroxyvalerate) and the blend PHB-PLA. The intensity of biopolymer hydrolysis characterized by total weight lost and the viscosity-averaged molecular weight (MW) decrement. The degradation is enhanced in the series PHBV<PHB<PHB-PLA blend PLA. Characterization of PHB and PHBV includes MW and crystallinity evolution (X-ray diffraction) and AFM analysis of PHB film surfaces before and after aggressive medium exposition. The important impact of MW on the biopolymer hydrolysis is shown.

The bacterial polyhydroxyalkanoates (PHAs) and their principal representative poly(3-R-hydroxybutyrate) (PHB) create a competitive option to conventional synthetic polymers such as polypropylene, polyethylene, polyesters et al. These polymers are nontoxic and renewable. Their biotechnology output does not depend on hydrocarbon production as well as their biodegradation intermediates and resulting products (water and carbon dioxide) do not provoke the adverse actions in environmental media or living systems [1–3]. Being environment friendly [4], the PHB and its derivatives are used as the alternative packaging materials, which are biodegradable in the soil or different humid media [5, 6].

The copolymerization of 3-hydroxybutyrate entities with 3-hydroxyoctanoate (HO), 3-hydroxyheptanoate (HH), or 3-hydroxyvalerate (HV) monomers modifies the physical and mechanical characteristics of the parent PHB, such as ductility and toughness to depress its processing temperature and embrittlement. Besides, copolymers PHB-HV [7], PHB-HH [8], or PHB-HO [9] et al., have improved thermophysical and/or mechanical properties, and hence they expand the spectrum of constructional and medical materials/items. For predicting the behavior of PHB and its copolymers in an aqueous media, e.g., *in vitro*, in a living body or in a wet soil, it is essential to study kinetics and mechanism of hydrolytic destruction.

Despite the history of such-like investigations reckons about 25 years, the problem of (bio)degradation in semicrystalline biopolymers is too far from a final resolution. Moreover, in the literature the description of hydrolytic degradation kinetics during long-term period is comparatively uncommon [10–14]. Therefore, the main object of this chapter is the comparison of long-term degradation kinetics for the PLA, PHB, and its derivatives, namely its copolymer with 3-oxyvalerate (PHBV) and the blend PHB/PLA.

The contrast between degradation profiles for PHB and PLA makes possible to compare the degradation behavior for two most prevalent biodegradable polymers. Besides, a significant attention is devoted to the impact of molecular weight (MW) for above polymer systems upon hydrolytic degradation and morphology (crystallinity and surface roughness) at physiological (37°C) and elevated (70°C) temperatures.

3.2 EXPERIMENTAL

3.2.1 MATERIALS

In this work, we have used poly-L-lactide (PLA) with different molecular weights: 67, 152, and 400 kDa (Fluka Germany); chloroform (ZAO EKOS-1, RF); sodium valerate (Sigma-Aldrich, USA); and monosubstituted sodium phosphate (NaH$_2$PO$_4$, ChimMed, RF).

3.2.2 PHAS PRODUCTION

The samples of PHB and copolymer of hydroxybutyrate and hydroxyvalerate (PHBV) have been produced in A.N. Bach's Institute of Biochemistry. A highly efficient strain-producer (80 wt. % PHB in the dry weight of cells), *Azotobacter chroococcum* 7Б, has been isolated from rhizosphere of wheat (the sod-podzol soil). Details of PHB biosynthesis have been published in Ref. [15]. Under conditions of PHBV synthesis, the sucrose concentration was decreased till 30 g/L in medium; and after 10 h incubation, 20 mm sodium valerate was added. Isolation and purification of the biopolymers were performed via centrifugation, washing, and drying at 60°C subsequently. Chloroform extraction of BPHB or BPHBV from the dry biomass and precipitation, filtration, washing again, and drying has been described in our previous work [15]. The monomer content (HB/HV ratio) in PHBV has been determined by nuclear magnetic resonance in accordance with procedure described earlier in Ref. [16]. The per cent concentration of HV moiety in the copolymer was calculated as the ratio between the integral intensity of methyl group of HV (0.89 ppm) and total integral intensity the same group and HB group (1.27 ppm). This value is 21 mol per cent.

3.2.3   MOLECULAR WEIGHT DETERMINATION

The viscosity-averaged MW was determined by the viscosity (η) measurement in chloroform solution at 30°C. The calculations of MW have been made in accordance with Mark–Houwink equation [17]:

$$[\eta] = 7.7 \cdot 10^{-5} \cdot M^{0.82}$$

3.2.4   *FILM PREPARATIONS OF PHAS, PLA, AND THEIR BLENDS*

The films of parent polymers (PHB, PHBV, and PLA) and their blends with the thickness about 40μm were cast on a fat-free glass surface. We obtained the set of films with different MW = 169 ± 9 (defined as PHB 170), 349 ± 12 (defined as PHB 350), 510 ± 15 kDa (defined as PHB 500), and 950 ± 25 kDa (defined as PHB 1,000) as well as the copolymer PHBV with MW = 1,056 ± 27 kDa (defined as PHBV).

In addition, we prepared the set of films on the base of PLA with same thickness 40 μm and MW = 67 (defined as PLA 70), MW = 150 and 400 kDa. Along with them, we obtained the blend PHB/PLA with weight ratio 1:1 and MW = 950 kDa for PHB, and MW = 67 kDa for PLA (defined as PHB + PLA blend). Both components mixed and dissolved in common solvent, chloroform, and then cast conventionally on the glass plate. All films were thoroughly vacuum processed for removing of solvent at 40°C.

3.2.5   HYDROLYTIC DEGRADATION *IN VITRO* EXPERIMENTS

Measurement of hydrolytic destruction of the PHB, PLA, PHBV films, and the PHB-PLA composite was performed as follows. The films were incubated in 15 ml 25 mm phosphate buffer, pH 7.4, at 37°C or 70°C in a ES 1/80 thermostat (SPU, Russia) for 91 days; pH was controlled using an Orion 420 + pH-meter (Thermo Electron Corporation, USA). For polymer weight measurements films were taken from the buffer solution every three day, dried, placed into a thermostat for 1 h at 40°C, and then weighed with a balance. The film samples weighed 50–70 mg each. The loss of polymer weight due to degradation was determined gravimetrically using an AL-64 balance (Acculab, USA). Every three days, the buffer was replaced by the fresh one.

3.2.6 WIDE ANGLE X-RAY DIFFRACTION

The PHB and PHBV chemical structure, the type of crystal lattice and crystallinity was analyzed by wide angle X-ray scattering (WAXS) technique. X-ray scattering study was performed on device on the basis of 12 kW generator with rotating copper anode RU-200 Rotaflex (Rigaku, Japan) using CuK radiation (wavelength $\lambda = 0.1542$ nm) operated at 40 kV and 140 mA. To obtain pictures of wide angle X-ray diffraction of polymers two-dimensional position-sensitive X-ray detector GADDS (Bruker AXS, Germany) with flat graphite monochromator installed on the primary beam was used. Collimator diameter was 0.5 mm [18]

3.2.7 ATOMIC FORCE MICROSCOPY OF PHB FILMS

Microphotographs of the surface of PHB films were obtained be means of atomic force microscopy (AFM). The AFM imaging was performed with Solver PRO-M (Zelenograd, Russia). For AFM imaging, a piece of the PHB film (~2 × 2 mm^2) was fixed on a sample holder by double-side adhesive tape. Silicon cantilevers NSG11 (NT-MDT, Russia) with typical spring constant of 5.1 N/m were used. The images were recorded in semi-contact mode, scanning frequency of 1–3 Hz, scanning areas from 3 × 3 to 20 × 20 µm^2, topography, and phase signals were captured during each scan. The images were captured with 512 × 512 pixels. Image processing was carried out using Image Analysis (NT-MDT, Russia) and FemtoScan Online (Advanced technologies center) software.

3.3   RESULTS AND DISCUSSION

The *in vitro* degradation of PHB with different MW and its derivatives (PHBV, blend PHB/PLA) prepared as films was observed by the changes of total weight loss, MW, and morphologies (AFM, XRD) during the period of 91 days.

3.3.1   THE HYDROLYSIS KINETICS OF PLA, PHB, AND ITS DERIVATIVES

The hydrolytic degradation of the biopolymer and the derivatives (the copolymer PHBV, and the blend PHB/PLA 1:1) has been monitored for 3 months under condition, which is realistically approximated to physiological conditions, namely, *in vitro*: phosphate buffer, Ph = 7.4, temperature 37°C. The analysis of kinetic curves for all samples shows that the highest rate of weight loss is observed for PLA with the smallest MW ≈ 70 kDa and for PHB with relatively low MW ≈ 150 kDa (Figure 3.1). On the base of the data in this figure, it is possible to compare the weight-loss increment for the polymers with different initial MW. Here, we clearly see that the samples with the higher MWs (300–1,000 kDa) are much stable against hydrolytic degradation than the samples of the lowest MW. The total weight of PHB films with MW = 150 kDa decreases faster compared to the weight reduction of the other PHB samples with higher MWs=300 and 450 or 1,000 kDa. In addition, by the 91st day of buffer exposition the residual weight of the low-MW sample reaches 10.5 per cent weight loss that it is essentially higher than the weight loss for the other PHB samples (see Figure 3.1 over again).

After establishing the impact of MW upon the hydrolysis, we have compared the weight-loss kinetic curves for PLA and PHB films with the relatively comparative MW = 400 and 350 kDa, respectively, and the same film thickness. For the PLA films one can see the weight depletion with the higher rate than the analogous samples of PHB. The results obtained here are in line with the preceding literature data [8, 12, 19–21].

Having compared the destruction behavior of the homopolymer PHB and the copolymer PHBV, we can see that the introduction of hydrophobic entity (HV) into the PHB molecule via copolymerization reveals the hydrolytic stability of PHBV molecules. For PHBV, a hydrolysis induction time is the longest among the other polymer systems and over a period of

70 days its weight loss is minimal (< 1% wt.) and possibly related with desorption of low-molecular fraction of PHBV presented initially in the samples after biosynthesis and isolation. The kinetic curves in Figure 3.1 show also that the conversion the parent polymers to their blend PHB-PLA decreases the hydrolysis rate compared with PHB (MW = 1,000 kDa) even if the second component is a readily hydrolysable polymer: PLA (MW = 70 kDa).

For the sake of hydrolysis amplification and its exploration simultaneously, a polymer exposition in aqueous media has usually been carried out at elevated temperature [11, 19]. To find out a temperature impact on degradation and intensify this process, we have elevated the temperature in phosphate buffer to 70°C. This value of temperature is often used as the standard in other publications (see e.g. Ref. [11]). As one should expect, under such condition, the hydrolysis acceleration is fairly visible that is presented in Figure 3.1(b). By the 45th day of PLA incubation, its films turned into fine-grinding dust with the weight-loss equaled 50 per cent (MW = 70 kDa) or 40 per cent (MW = 350 kDa). Simultaneously, the PHB with the lowest MW = 170 kDa has the weight loss=38 wt per cent and the film was markedly fragmented while the PHB samples with higher MWs 350; 500; and 1,000 kDa have lost the less per cent of the initial weight, namely 20, 15, and 10 per cent, respectively. In addition, for 83 days, the weight drop in the PHB-PLA blend films is about 51 wt. per cent and, hence, hydrolytic stability of the blend polymer system is essentially declined (see. Figures 3.1(a) and 3.1(b)).

At elevated temperature of polymer hydrolysis (70°C) as well as at physiological temperature 37°C, we have demonstrated again that the PHBV films are stable because by 95th day, they lost only 4 wt per cent. The enhanced stability of PHBV relative to the PHB has been confirmed by other literature data [21]. Here, it is worth remarking that during biosynthesis of the PHBV, two opposite effects of water sorption acting reversely each other occur. On the one hand, while the methyl groups are replaced by ethyl groups, the total hydrophobicity of the copolymer is enhanced, on the other hand; this replacement leads to decrease of crystallinity in the copolymer [22]. The interplay between two processes determines a total water concentration in the copolymer and hence the rate of hydrolytic degradation. In general, in the case of PHBV copolymer (HB/HV = 4:1 mol ratio), the hydrophobization of its chain predominates the effect of crystallinity decrease from 75 per cent for PHB to ~60 per cent for PHBV.

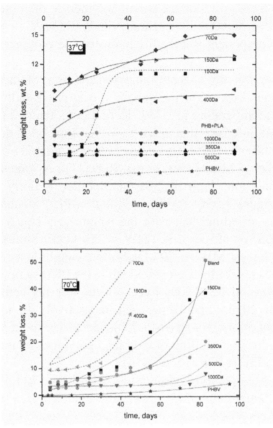

FIGURE 3.1 Weight loss in the phosphate buffer for PHB and its derivatives with different MW (shown on the curves in kDa). 37°C, 70°C: ♦, ▶, and ◀ are PLA films with MW = 70, 150, and 400 kDa respectively; ■, ▲, ●, and ▼ are PHB samples with 170; 350; 500; and 1,000 kDa, respectively; PHBV 1050 (★); and PHB-PLA blend (●).

3.3.2 CHANGE OF MOLECULAR WEIGHT FOR PHB AND PHBV

On exposure of PHB and PHBV films to buffer medium at physiological (37°C) or elevated (70°C) temperatures, we have measured both their total weight loss and the change of their MW simultaneously. In particular, we have shown the temperature impact on the MW decrease that will be much clear if we compare the MW decrements for the samples at 37° and 70°C. At 70°C, the above biopolymers have a more intensive reduction of MW compared with the reduction at 37°C (see Figure 3.2). In particular, at elevated temperature, the initial MW (= 350 kDa) has the decrement by seven times more than the MW decrement at physiological condition. In general, the final MW loss is nearly proportional to the initial MW of sample that is correct especially at 70°C. As an example, after the 83-days incubation of PHB films, the initial MW = 170 kDa dropped as much as 18 wt per cent and the initial MW = 350 kDa has the 9.1 wt per cent decrease.

Figure 3.2 shows that the sharp reduction of MW takes place for the first 45 days of incubation; and after this time, the MW change becomes slow. Combining the weight-loss (Section 3.1) and the MW depletion, it is possible to present the biopolymer hydrolysis as the two-stage process. On the initial stage, the random cleavage of macromolecules and the MW decrease without a significant weight-loss occur. Within this time, the mean length of PHB intermediates is fairly large and the molar ratio of the terminal hydrophilic groups to the basic functional groups in a biodegradable fragment is too small to provide the solubility in aqueous media. This situation is true for the PHB samples with middle and high MW (350; 500; and 1,000 kDa) when at 37°C their total weight remains stable during all time of observation, but the MW values are decreased until 76, 61, and 51 wt per cent, respectively.

On the second stage of degradation, when the MW of the intermediate molecules attains the some "critical" value and the products of hydrolysis become hydrophilic to provide dissolution and diffusion into water medium, the weight reduction is clearly observed at 70°C. This stage is accompanied by the changes of physical–chemical, mechanical, and structural characteristics and a geometry alteration. A similar two-stage mechanism of PHB degradation has been described in the other publications [23, 24]. Further, in the classical work of Reush [25], she showed that hydrophilization of PHB intermediates occurs at relatively low MW namely, at several decades of kDa. Our results provide evidences that the reduction of MW till "critical" values to be equal about 30 kDa leads to the expansion of the second stage, namely, to the intensive weight loss.

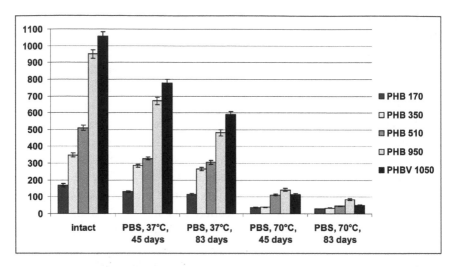

FIGURE 3.2 The molecular weight conversion of PHB and PHBV films during hydrolysis in phosphate buffer (PBS), pH = 7.4°C, 37°C, and 70°C.

3.3.3 CRYSTALLINITY OF PHB AND PHBV

We have above revealed that during hydrolytic degradation, PHB and PHBV show the MW reduction (Section 3.2) and the total weight decrease (Section 3.1). In addition, by the X-ray diffraction (XRD) technique, we have measured the crystallinity degree of PHB and PHBV that varied depending on time in the interval of values 60–80 per cent (see Figure 3.3(a)). We have noted that on the initial stage of polymer exposition to the aqueous buffer solution (at 37°C for 45 days), the crystallinity degree has slightly increased and then, under following exposition to the buffer, this characteristic is constant or even slightly decreased showing a weak maximum.

When taking into account that at 37°C, the total weight for the PHB films with MWs equal 350; 500; and 1,000 kDa, and the PHBV film with MW equals 1,050 are invariable, a possible reason of the small increase in crystallinity is recrystallization described earlier for PLA [26]. Recrystallization (or additional crystallization) happens in semicrystalline polymers where the crystallite portion can increase using polymer chains in adjoining amorphous phase [22].

At higher temperature of hydrolysis, 70°C, the crystallinity increment is strongly marked and has a progressive trend. The plausible explanation of this effect includes the hydrolysis progress in amorphous area of biopolymers. It is well known that the matrices of PHB and PHBV are formed by alternative crystalline and noncrystalline regions, which determine both polymer morphologies and transport of aggressive medium. In addition, we have revealed recently by H–D exchange FTIR technique that the functional groups in the PHB crystallites are practically not accessible to water attack.

Therefore, the hydrolytic destruction and the weight decrease are predominantly developed in the amorphous part of polymer [22, 27]. Hence, the crystalline fraction becomes larger through polymer fragment desorption from amorphous phase. This effect takes place under the strong aggressive conditions (70°C) and does not appear under the physiological conditions (37°C) when the samples have invariable weight.

Owing to the longer lateral chains in PHBV, copolymerization modifies essentially the parent characteristics of PHB such as decreasing in crystallinity, the depression of melting and glass temperatures and, hence, enhancing ductility and improvement of processing characteristics [14, 28, 29]. In addition, we have founded out that the initial crystallinity of PHB films is a monotonically increased function of initial MW (see Figure 3.3(b)). For samples with relatively low molecular weight, it is difficult to compose the perfect crystalline entities because of a relatively high concentration of terminal groups performing as crystalline defects.

Thus, at physiological temperature the crystallinity, measured during degradation by XRD technique, has an slightly extreme character. On the initial stage of PHB degradation, the crystalline/amorphous ratio is increased owing to additional crystallization through involvement of polymer molecules situated in amorphous fields. By contrast, at 70°C after reaching the critical MW values (see Section 3.2), the following desorption of water-soluble intermediates occurs. On the following stage, as the degradation is developed till film disintegration, the crystallinity drop must take place as result of crystallite disruption.

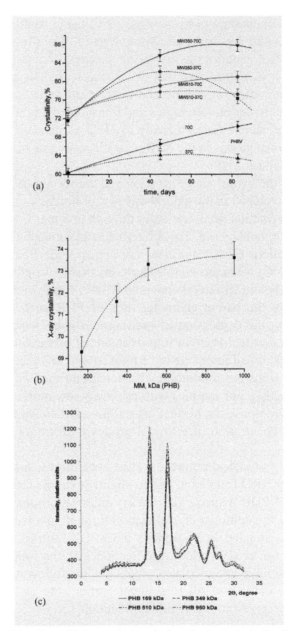

FIGURE 3.3 (a)—Crystallinity evolution during the hydrolysis for PHB and PHBV films (denoted values of temperature and MW). (b)—crystallinity as function of initial MW for PHB films prepared by cast method, (c) X-ray diffractograms for PHB films with different molecular weight given under X-axis.

3.3.4   THE ANALYSIS OF FILM SURFACES FOR PHB BY AFM TECHNIQUE

Morphology and surface roughness of PHB film exposed to corrosive medium (phosphate buffer) have been studied by the AFM technique. This experiment is important for surface characterization because the state of implant surface determines not only mechanism of degradation but also the protein adsorption and cell adhesion that are responsible for polymer biocompatibility [30]. As the standard sample, we have used the PHB film with relatively low MW = 170 kDa.

The film casting procedure may lead to distinction in morphology between two surfaces when the one plane of the polymer film was adjacent with glass plate and the other one was exposed to air. In reality, as it is shown in Figure 3.4, the surface exposed to air has a roughness formed by a plenty of pores with the length of 500–700 nm. The opposite side of the film contacted with glass (Figure 3.4(b)) is characterized by minor texture and by the pores with the less length as small as 100 nm. At higher magnification (here not presented) in certain localities, it can see the stacks of polymer crystallites with width about 100 nm and length 500–800 nm.

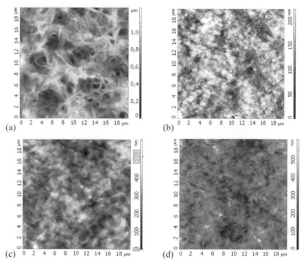

FIGURE 3.4   AFM topographic images of PHB films (170 kDa) with a scan size of 18 × 18 μm: the rough surface of fresh-prepared sample (exposed to air)—(a) the smooth surface of fresh-prepared sample (exposed to glass)—(b) the sample exposed to phosphate buffer at 37°C for 83 days—(c) the sample exposed to phosphate buffer at 70°C for 83 days—(d) general magnificence is 300.

Inequality of morphology between two surfaces gets clearly evident when quantitative parameters of roughness (r_n) were compared. A roughness analysis has shown that averaged value of this characteristic

$$R_a = \frac{1}{N} \sum_{n=1}^{N} |r_n|$$

and a root-mean-square roughness

$$R_q = \sqrt{\frac{1}{N} \sum_{n=1}^{N} r_n^2}$$

for surfaces exposed to glass or air differ about ten times.

The variance of characteristics is related with solvent desorption conditions during its evaporation for the cast film. During chloroform evaporation from the surface faced to air, the flux forms additional channels (namely the pores), which are fixed as far as the film is solidified and crystallized. Simultaneously, during evaporation the morphology and texture on the opposite side of film exposed to the glass support are not subjected to the impact of solvent transport.

The morphology of the latter surface depends on energy interaction conditions (interface glass-biopolymer tension) predominantly. The exposition of PHB films to the buffer for long time (83 days) lead to a threefold growth of roughness characteristics for glass-exposed surface and practically does not affects the air-exposed surface. It is interesting that temperature of film degradation does not influence on the roughness change. The surface characteristics of film surface have the same values after treatment at 37°C and 70°C.

Summarizing the AMF data, we can conclude that during degradation the air-exposed, rough surface remained stable that probably related with the volume mechanism of degradation (V-mechanism [31, 32]). The pores on the surface provide the fast water diffusion into the bulk of PHB. However, under the same environmental conditions, the change of surface porosity (roughness) for glass-exposed surface is remarkable, showing the engagement of surface into degradation process (S-mechanism [31, 32]).

Past findings show that along with the volume processes of polymer degradation the surface hydrolysis can proceed. Several authors [20, 21] have recently reported on surface mechanism of PHB destruction, but traditional viewpoint states a volume mechanism of degradation [12]. Here, using an advanced method of surface investigation (AMF), we have shown that for the same film under the same exterior conditions, the mechanism of degradation could be changed depending on the prehistory of polymer preparation.

3.4   CONCLUSION

Analyzing all results related with hydrolytic degradation of PHB and its derivatives, the consecutive stages of such complicated process are presented as follows. During the initial stage, the total weight is invariable and the cleavage of biomolecules resulting in the MW decrease is observed. Within this time, the PHB intermediates are too large and hydrophobic to provide solubility in aqueous media. Because the PHB crystallites stay stable, the crystallinity degree is constant as well and even it may grow up due to additional crystallization. On the second stage of hydrolysis, when the MW of intermediates attain the "critical" value, which is equal about 30 kDa, these intermediates can dissolve and diffuse from the polymer into buffer. Within this period, the weight loss is clearly observed. The intensity of hydrolysis characterized by the weight loss and the MW decrement is enhanced in the series PHBV<PHB<PHB-PLA blend<PLA.

The growth of initial MW (a terminal group reducing) impacts on the hydrolysis stability probably due to the increase of crystallite perfection and crystallinity degree. The XRD data reflect this trend (see Figure 3.3(b)). Moreover, the surface state of PHB films explored by AFM technique depends on the condition of film preparation. After cast processing, there is a great difference in morphologies of PHB film surfaces exposed to air and to glass plate.

It is well known that the mechanism of hydrolysis could include two consecutive processes: 1) volume degradation and 2) surface degradation. Under essential pore formation (in the surface layer exposed to air), the volume mechanism prevails. The smooth surface of PHB film contacted during preparation with the glass plate is degraded much intensely than the opposite rough surface (Figure 3.4).

In conclusion, we have revealed that the biopolymer MW determines the form of a hydrolysis profile (see Figure 3.1). For acceleration of this process, we have to use the small MW values of PHB. In this case, we affect both the degradation rate and the crystalline degree (Figure 3.1b). By contrast, for prolongation of service-time in a living system, it is preferable to use the high-MW PHB that is the most stable polymer against hydrolytic degradation.

KEYWORDS

- atomic force microscopy
- Biodegradable polymers
- Crystallinity
- X-ray diffraction

REFERENCES

1. Sudesh, K.; Abe, H.; and Doi, Y.; Synthesis, structure and properties of polyhydroxy-alkanoates: Biological polyesters. *Progress Polym. Sci. (Oxford).* **2000**, *25(10)*, 1503–1555.
2. Lenz, R. W.; and Marchessault, R. H.; Bacterial polyesters: biosynthesis, biodegradable plastics and biotechnology. *Biomacromole.* **2005**, *6(1)*, 1–8.
3. Bonartsev, A. P.; Iordanskii, A. L.; Bonartseva, G. A.; and Zaikov, G. E.; Biodegradation and medical application of microbial poly (3-hydroxybutyrate). *Polym. Res. J.* **2008**, *2(2)*, 127–160.
4. Kadouri, D.; Jurkevitch, E.; Okon, Y.; and Castro-Sowinski, S.; Critical reviews in microbiology. *Ecol. Agricul. Significance Bacterial Polyhydroxyalkanoates.* **2005**, *31(2)*, 55–67.
5. Jendrossek, D.; and Handrick, R.; Microbial degradation of polyhydroxyalkanoates. *Ann. Rev. Microbiol.* **2002**; *56*, 403–432.
6. Steinbuchel, A.; and Lutke-Eversloh, T.; Metabolic engineering and pathway construction for biotechnological production of relevant polyhydroxyalkanoates in microorganisms. *Biochem. Eng. J.* **2003**; *16*, 81–96.
7. Miller, N. D.; and Williams, D. F.; On the biodegradation of poly-beta-hydroxybutyrate (PHB) homopolymer and poly-beta-hydroxybutyrate-hydroxyvalerate copolymers. *Biomater.* **1987**, *8(2)*, 129–137.

8. Qu, XH.; Wu, Q.; Zhang, KY.; and Chen, GQ.; In vivo studies of poly(3-hydroxybutyrate-co-3-hydroxyhexanoate) based polymers: biodegradation and tissue reactions. *Biomater.* **2006***; 27(19)*, 3540–3548.

9. L.J.R. Fostera, V. Sanguanchaipaiwonga, C.L. Gabelisha, J. Hookc, M. Stenzel. A natural-synthetic hybrid copolymer of polyhydroxyoctanoate-diethylene glycol: biosynthesis and properties. *Polym.* **2005,** *46,* 6587–6594.

10. Marois, Y.; Zhang, Z.; Vert, M.; Deng, X.; Lenz, R.; and Guidoin, R.; Mechanism and rate of degradation of polyhydroxyoctanoate films in aqueous media: a long-term in vitro study. *J. Biomed. Mater. Res.* **2000***; 49(2)*, 216–224.

11. Freier, T.; et al. In vitro and in vivo degradation studies for development of a biodegradable patch based on poly(3-hydroxybutyrate). *Biomater.* **2002,** *23,* 2649–2657.

12. Doi, Y.; Kanesawa, Y.; Kawaguchi, Y.; Kunioka, M.; Hydrolytic degradation of microbial poly(hydroxyalkanoates). *Makrom Chem. Rapid Commun.* **1989,** *10,* 227–230.

13. Renstadt, R.; Karlsson, S.; Albertsson, A. C.; The influence of processing conditions on the properties and the degradation of poly(3-hydroxybutyrate-co-3-hydroxyvalerate). *Macromol. Symp.* **1998,** *127,* 241–249.

14. Cheng Mei-Ling, Chen Po-Ya, Lan Chin-Hung, Sun Yi-Ming, Structure, mechanical properties and degradation behaviors of the electrospun fibrous blends of PHBHHx/PDLLA. *Polym.* **2011,** *doi:10.1016/j.polymer.2011.01.039 in press.*

15. Myshkina, V.L.; Nikolaeva, D. A.; Makhina, T. K.; Bonartsev, A. P.; and Bonartseva G.A. Effect of Growth conditions on the Molecular weight of poly-3-hydroxybutyrate produced by Azotobacter chroococcum 7B. *Appl. Biochem. Microbiol.* **2008,** *44(5),* 482–486.

16. Myshkina, V. L.; et al. Biosynthesis of poly-3-hydroxybutyrate-3-hydroxyvalerate copolymer by Azotobacter chroococcum strain 7B. *Appl. Biochem. Microbiol.* **2010,** *46(3),* 289–296.

17. Akita, S.; Einaga, Y.; Miyaki, Y.; and Fujita, H.; Solution Properties of Poly(D-β-hydroxybutyrate). 1. Biosynthesis and Characterization. *Macromole.* **1976,** *9,* 774–780.

18. Rebrov, A.V.; Dubinskii, V. A.; Nekrasov, Y. P.; Bonartseva, G. A.; Shtamm, M.; and Antipov, E. M.; [Structure phenomena at elastic deformation of highly oriented polyhydroxybutyrate. *Polym. Sci. (Russian).* **2002,** *44A,* 347–351.

19. Koyama, N.; and Doi, Y.; Morphology and biodegradability of a binary blend of poly((R)-3-hydroxybutyric acid) and poly((R,S)-lactic acid). *Can. J. Microbiol.* **1995,** *41(Suppl. 1),* 316–322.

20. Majid, M. I. A.; Ismail, J.; Few, L. L.; and Tan, C. F.; The degradation kinetics of poly(3-hydroxybutyrate) under non-aqueous and aqueous conditions. *Euro. Polym. J.* **2002,** *38(4),* 837–839.

21. Choi, G. G.; Kim, H. W.; and Rhee, Y. H.; Enzymatic and non-enzymatic degradation of poly(3-hydroxybutyrate-co-3-hydroxyvalerate) copolyesters produced by Alcaligenes sp. MT-16. *J. Microbiol.* **2004,** *42(4),* 346–352.

22. Iordanskii, A. L.; Rudakova, T. E.; Zaikov, G. E.; Interaction of polymers with corrosive and bioactive media. New York–Tokyo: VSP; **1984.**

23. Wang, H. T.; Palmer, H.; Linhardt, R. J.; Flanagan, D. R.; and Schmitt, E.; Degradation of poly(ester) microspheres. *Biomater.* **1990,** *11(9),* 679–685.

24. Kurcok, P.; Kowalczuk, M.; Adamus, G.; Jedlinrski, Z.; and Lenz, R. W.; Degradability of poly (b-hydroxybutyrate)s. Correlation with chemical microstucture. *JMS-Pure Appl. Chem.* **1995**, *A32*, 875–880.

25. Reusch, R. N.; Biological complexes of poly-β-hydroxybutyrate. FEMS *Microbiol. Rev.* **1992**, *103*, 119–130.

26. Molnár, K.; Móczó, J.; Murariu, M.; Dubois, Ph.; and Pukánszky, B.; Factors affecting the properties of PLA/CaSO4 composites: homogeneity and interactions. *Express Polym. Lett.* **2009**, *3(1)*, 49–61.

27. Spyros, A.; Kimmich, R.; Briese, B.; and Jendrossek, D.; 1H NMR imaging study of enzymatic degradation in poly(3-hydroxybutyrate) and poly(3-hydroxybutyrate-co-3-hydroxyvalerate). Evidence for preferential degradation of amorphous phase by PHB depolymerase B from Pseudomonas lemoignei. *Macromole.* **1997,** *(30)*, 8218–8225.

28. Luizier, W. D.; Materials derived from biomass/biodegradable materials. *Proc. Natl. Acad. Sci. USA.* **1992**, *89*, 839–842.

29. Gao, Y.; et al. *Eur. Polym. J.* **2006**, *42(4)*, 764–75.

30. Pompe, T.; et al. Surface modification of poly(hydroxybutyrate) films to control cell-matrix adhesion. *Biomater.* **2007**, *28(1)*, 28–37.

31. Siepmann, J.; Siepmann, F.; and Florence, A. T.; Local controlled drug delivery to the brain: Mathematical modeling of the underlying mass transport mechanisms. *Int. J. Pharmaceutics.* **2006**, *314(2)*, 101–119.

32. Zhang, T.C.; Fu, Y. C.; Bishop, P. L.; et al. Transport and biodegradation of toxic organics in biofilms. *J. Hazardous Mater.* **1995**, *41(2–3)*, 267–285.

CHAPTER 4

A DETAILED REVIEW ON PHYSICOCHEMICAL PROPERTIES, SYNTHESIS, AND APPLICATION OF POLYACETYLENE

O. A. PONOMAREV, A. I. RAKHIMOV, N. A. RAKHIMOV,
E. S. TITOVA, V. A. BABKIN, and G. E. ZAIKOV

CONTENTS

4.1 INTRODUCTION

Features of synthesis, structure, properties, and use of polyacetylenes are considered in this chapter. The catalytic polymerization of acetylene using different catalysts is shown. Plasmachemical synthesis of carbines is considered. The results of studying the structure of polyacetylenes by electron spectroscopy are presented. The results of the research of the surface morphology of polyacetylene are presented. Agency of receiving methods of polyacetylene on its properties is shown.

4.2 SYNTHESIS AND STRUCTURE OF POLYACETYLENE

The chemical element in periodic system of D.I. Mendeleev—carbon possesses a variety of unique propertes. Scientists thought that two forms exists of crystal carbon only-diamond and graphite (opinion beginning of 60th years of XX century). These forms are widespread in the nature, and they are known to mankind with the most ancient times.

The question on an opportunity of existence of forms of carbon with sp-hybridization of atoms was repeatedly considered theoretically. In 1885, German chemist A. Bayer tried to synthesize chained carbon from derivatives of acetylene by a step method. However, Bayer's attempt to receive polyin has appeared unsuccessful. He received the hydrocarbon consisting from four molecules of acetylene, associated in a chain, and appeared extremely unstable.

A. M. Sladkov, V. V. Korshak, V. I. Kasatochkin, and Yu. P. Kudryavcev [1] observed loss of a black sediment of polyin compound of carbon having the linear form at transmission acetylene in a water—ammonia solution of salt Cu (II) (oxidizing dehydropoly condensation of acetylene led obviously to polyacetylenides of copper). This powder blew up at heating in a dry condition, and in damp—at a detonation. Process of oxidative dehydropoly condensation of acetylene can be written down in a following kind schematically [1] at $x+y+z=n$:

$$n \ H-C\equiv C-H$$

$$\Big\downarrow Cu^{2+}$$

$$H(-C\equiv C-)_x Cu \ + \ H(-C\equiv C-)_y H \ + \ Cu(-C\equiv C-)_z H$$

$$\Big\downarrow FeCl_3$$

$$H(-C\equiv C-)-H$$
$$n$$

At surplus of ions Cu^{2+}, the mix of various polyins and polyacetyl-enides of copper various molecular weight is formed. Additional oxidation of products received at this stage (with help $FeCl_3$ or $K_3[Fe(CN)_6]$) leads to formation polyins with the double molecular weight. The last do not blow up any more at heating and impact, but contain a plenty of copper. Possibly, trailer atoms of copper stabilize polyins by dint of to complex-ation.

The content by carbon was 90 per cent of clean polyin (cleaning cleared from copper and impurity of other components of the reactionary medi-um). Only multihours' heating of samples of polyin at 1,000°C in vacuum has allowed to receive analytically pure samples of α-carbyne. Similar processing results not only in purification but also in partial crystallization of polyacetylene.

Under A.M. Sladkov's offer such polyacetylene have named «carbyne» * (from Latin *carboneum* (carbon) with the termination «in», accepted in organic chemistry for a designation of acetylene bond).

By acknowledgement of polyin structures in chains is formation of oxalic acid after ozonation hydrolysis of carbine [2, 3]:

$$(-C\equiv C-)_n \ \xrightarrow{O_3} \ \left(\begin{array}{c} -C\equiv C- \\ | \ \ | \\ O \ \ O \\ \diagdown\!\diagup \\ O \end{array}\right)_n \ \xrightarrow{H_2O} \ n HOOC-COOH.$$

New linear polymer with cumulene bonds was received [2, 3]. It is named polycumulene. The proof of such structure became the fact that at ozonation of polycumulene, only carbon dioxide is received:

$$(=C=C=)_n \xrightarrow{O_3} 2nCO_2.$$

Cumulene modification of carbyne (β-c arbyne) has been received on specially developed by Sladkov two-stage method [3]. At the first stage, spent polycondensation of suboxide of carbon (C_3O_2) with dimagnesium dibromine acetylene as Grignard reaction with formation polymeric glycol:

$$nO=C=C=C=O + nBrMgC\equiv CMgBr \rightarrow \left(-C\equiv C-\underset{OH}{C}=C=\underset{OH}{C}-\right)_n.$$

At the second stage, this polymeric glycol reduced by stannous chloride hydrochloric acid:

$$\left(-C\equiv C-\underset{OH}{C}=C=\underset{OH}{C}-\right)_n \xrightarrow[-(HCl + SnO_2)]{+ SnCl_2} (=C=C=C=C=C=)_n.$$

High-molecular cumulene represents an insoluble dark-brown powder with the developed specific surface (200–300 m^2/g) and density 2.25 g/sm^3. At multihours heating at 1,000°C and the depressed pressure, polycumulene partially crystallizes. Two types of monocrystals have been found out in received after annealing a product by means of transmission electronic microscopy. Crystals corresponded to α- and β-modifications of carbyne.

One of the most convenient and accessible methods of reception carbyne or its fragments is reaction of dehydrohalogenation of the some polymers content of halogens (GP). Feature of this method is formation of the carbon chain at polymerization corresponding monomers. The problem at synthesis carbyne consists only in that at full eliminating of halogen hydride with formation of linear carbon chain. Exhaustive dehydrohalogenation is possible, if the next atoms of carbon have equal quantities of atoms of halogen and hydrogen. Therefore convenient GP for reception of carbyne were various polyvinyliden halogenides (bromides, chlorides and fluorides), poly(1,2-dibromoethylene), poly (1,1,2 or 1,2,3-trichlorobutadiene), for example:

$$(-CH_2-CHal_2-)_n \xrightarrow[-nHHal]{+B^-} (-CH=CHal-)_n \xrightarrow[-nHHal]{+B^-} (=C=C=)_n$$

The reaction of dehydrohalogenation typically carry out at presence of solutions of alkalis (B^-) in ethanol with addition of polar solvents. At use of tetrahydrofuran synthesis goes at a room temperature. This method allows avoiding course of collateral reactions. The amorphous phase only cumulene modification of carbyne is received as a result. Then, crystal of β-carbyne is synthesized from amorphous carbine by solid-phase crystallization.

Next method is dehydrogenation of polyacetylene. At interaction of polyacetylene with metallic potassium at 800°C and pressure 4 GPa led to dehydrogenation and formation of potassium hydride, the carbon matrix containing potassium. After removal of potassium from products (acid processing) precipitate out brown plate crystals of β-carbyne in hexagonal forms by diameter ~1 mm and thickness up to 1 mm. Carbyne also can be received by various methods of chemical sedimentation from a gas phase.

Plasmochemical Synthesis of Carbyne. At thermal decomposition of hydrocarbons (acetylene, propane, heptane, benzol), carbon tetrachloride, carbon bisulfide, acetone in a stream nitrogen plasma is received the disperse carbon powders containing carbyne. Monocrystals of white color and (white or brown) polycrystals remain after selective oxidation of aromatic hydrocarbon. It is positioned, that formation of carbyne does not depend by nature initial organic compound. The moderate temperature of plasma (~3,200 K) and small concentration of reagents promote process.

Laser sublimations of carbon. Carbyne has been received at sedimentation on a substrate of steams of negative ions of the carbon after laser evaporation of graphite in 1971. The silvery-white layer was received on a substrate. This layer, according to data X-ray and diffraction researches, consists of amorphous and crystal particles of carbyne with the average size of crystallites $>10^{-5}$ sm.

Arc cracking of carbon. Evaporation in electric arc spectrally pure coals with enough slow polymerization and crystallization of a carbon steam on a surface of a cold substrate yields to product in which prevail carbyne forms of carbon.

Ion-stimulated precipitation of carbyne. At ion-stimulated condensation of carbon on a lining simultaneously or alternately the stream of carbon and a stream of ions of an inert gas moves. The stream of carbon is received by thermal or ionic evaporation of graphite. This method allows to receive carbyne films with a different degree of orderliness (from amorphous up to monocrystalline layers), carbynes of the set updating, and also a film of other forms of carbon. Annealing of films of amorphous carbon with various near order leads to crystallization of various allotropic forms carbon, including carbyne.

Sladkov has drawn the following conclusions on the basis of results of experiments on synthesis carbyne by methods of chemical sedimentation from gas phase:

- White sediments of carbyne are received, possibly, in the softest conditions of condensation of carbon: high enough vacuum, small intensity of a stream, and low energy of flying atoms or groups of atoms, small speed of sedimentation.
- Chains, apparently, grow perpendicularly to a lining, not being crosslinked among themselves.
- Probably, being an environment monovalent heteroatoms stabilise chains, do not allow them to be crosslinked.

Reception of carbyne from carbon graphite materials leads by heating of cores from pyrolytic graphite at temperature 2,700–3,200 K in argon medium. This leads to occurrence on the ends of cores a silvery-white strike (already through 15–20 s). This strike consists of crystals carbyne that is confirmed by data of method electron diffraction.

In 1958 Natta with employees are polymerized acetylene on catalyst system $Al(C_2H_5))_3$—$Ti(OC_3H_7)_4$ [7, 8]. The subsequent researches [9–11] led to reception of films stereoregular polyacetylene. The catalyst system $Al(Et)_3$—$Ti(OBu)_4$ provides reception of films of polyacetylene predominantly (up to 98%).

Films of polyacetylene are formed on a surface of the catalyst or practically to any lining moistened by a solution of the catalyst (it is preferable in toluene), in an atmosphere of the cleared acetylene [11]. The temperature and pressure of acetylene control growth of films [12, 13]. Homogeneous catalyst system before use typically maintain at a room temperature. Thus, reactions of maturing of the catalyst occur [14]:

$$Ti(OBu)_4 + AlEt_3 \rightarrow EtTi(OBu)_3 \rightarrow AlEt_2(OBu)$$

$$2EtTi(OBu)_3 \rightarrow 2Ti(OBu)_3 + CH_4 + C_2H_6$$

$$Ti(OBu)_3+AlEt_3 \rightarrow EtTi(OBu)_2+Al(Et)_2OBu$$

$$EtTi(OBu)_2+AlEt_3 \rightarrow EtTi(OBu)_2+Al(Et)_3$$

Aging of the catalyst in the beginning raises its activity. However, eventually the yield of polyacetylene falls because of the further reduction of the titan:

$$EtTi(OBu)_3+Al(Et)_3 \rightarrow Ti(OBu)_4+Al(Et)_2(OBu)+C_2H_4+C_2H_6$$

The jelly-like product of red color is formed if synthesis led at low concentration of the catalyst. This product consists from confused fibrils in the size up to 800 Å. Foam material with density from 0.4 to 0.4 g/sm^3 is possible to receive from the dilute gels by sublimation of solvent at temperature below temperature of its freezing [15].

Research of speed dependence for formation of films of polyacetylene from catalyst concentration and pressure of acetylene has allowed to find an optimum parity of components for catalyst Al/Ti = 4. Increase of this parity up to 10 leads to increase in the sizes of fibrils [16]. Falling of speed of reaction in the end of process speaks deterioration of diffusion a monomer through a layer of the film formed on a surface of the catalyst [17]. During synthesis, the film is formed simultaneously on walls of a flask and on a surface a catalytic solution. Gel collects in a cortex. The powder of polyacetylene settles at the bottom of a reactor. The molecular weight (M_n) a powder below, than gel, is also depressed with growth of concentration of the catalyst up to 400–500 [18]. The molecular weight of jellous polyacetylene slightly decreases with growth of concentration of the catalyst and grows from 2×10^4 up to $3, 6 \times 10^4$ at rise in temperature from −78 up to −10°C. The molecular weight of polyacetylene in a film is twice less, than in gel [19].

Greater sensitivity of the catalyst to impurity does not allow estimating unequivocally influence of various factors on M_p of polyacetylene. Low-molecular products with $M_p \sim 1,200$ are formed at carrying out of synthesis in the medium of hydrogen [20]. Concentration of acetylene renders significant influence on M_p: at increase of its pressure up to 760 mm Hg increases M_p up to 1,20,000 [14, 21]. Apparently, the heterogeneity of a substratum arising because of imperfection of technics for synthesis, is the reason for some irreproducibility properties of the received polymers. It is supposed that synthesis is carried out on "surface" of catalytic clusters. Research by EPR method has allowed to distinguish in the catalyst up to four types of complexes [22]. Polyacetylene cis-transoid structures is formed as a result.

Formation of trans-structure at heats speaks thermal isomerization. The alternative opportunity trans-disclosing of triple bond in a catalytic complex for a transitive condition is forbidden spatially [22, 23]. The structure of a complex and a kinetics of polymerization are considered in works in more detail [14, 24].

Parshakov A. S. with coauthors [25] have offered a new method of synthesis organo-inorganic composites nanoclusters transitive metals in an organic matrix by reactions of compounds of transitive metals of the maximum degrees of oxidation with monomers that at the first stage represent itself as a reducer. Formed thus clusters metals of the lowest degrees of oxidation are used for catalysis of polymerization a monomer with formation of an organic matrix.

Thus it is positioned, that at interaction $MoCl_5$ with acetylene in not polar mediums there is allocation HCl, downturn of a degree of oxidation of molybdenum and formation metallo-organic nanoclusters. Two distances Mo–Mo are found out in these nanoclusters by method EXAFS spectroscopy. In coordination sphere, Mo there are two nonequivalent atom of chlorine and atom of carbon. On the basis of results MALDI-TOF mass spectrometry, the conclusion is made that cluster of molybdenum has 12- or the 13-nuclear metal skeleton and its structure can be expressed by formulas $[Mo_{12}Cl_{24}(C_{20}H_{21})]^-$ or $[Mo_{13}Cl_{24}(C_{13}H_8)]^-$.

Greater sensitivity of the catalyst impurity does not allow to estimate unequivocally influence of various factors on M_n of polyacetylene. Low-molecular products with $M{\sim}1,200$ are formed at carrying out of synthesis in the environment of hydrogen [20]. Concentration of acetylene renders significant influence on $M_n{:}M_n$ increases up to 1,20,000 at increase of its pressure up to 760 mm hg [14, 21]. Apparently, the heterogeneity of a substratum arising because of imperfection of technics of synthesis, is the reason for some non repeatability properties of the received polymers. It is supposed that synthesis is carried out on "surface" of catalytic clusters. Research by method EPR has allowed to distinguish in the catalyst up to 4 types of complexes [22].

A Polyacetylene cis-transoid structures is as a result formed. Formation a trans-structure at heats speaks thermal isomerization, previous crystallization of circuits. The alternative opportunity a trans-disclosing of triple bondin in a catalytic complex for a transitive condition is forbidden spatially [22, 23]; in more detail, the structure of a complex and a kinetics of polymerization are considered in works [14, 24].

Parshakov A.S. with coauthors [25] have offered a new method of synthesis of organo-inorganic composites nanoclusters of transitive metals in an organic matrix by reactions of connections of transition metals of the maximum degrees of oxidation with monomers. Monomers represent itself as a reducer at the first stage. Formed thus clusters of metals of the lowest degrees of oxidation are used for a catalysis of polymerization a monomer with formation of an organic matrix.

Thus it is positioned, that at interaction $MoCl_5$ with acetylene in not polar mediums there is allocation HCl, downturn of a degree of oxidation of molybdenum and formation metallo-organic nanoclusters. Two distances Mo-Mo are found out in these clusters by a method of EXAFS-spectroscopy. Two nonequivalent atoms of chlorine and atom of carbon are available in coordination sphere Mo. The conclusion is made on the basis of results MALDI-TOF of mass-spectrometry, that cluster molybdenum has 12 or the 13-nuclear metal skeleton and its structure can be expressed by formulas $[Mo_{12}Cl_{24}(C_{20}H_{21})]^-$ or $[Mo_{13}Cl_{24}(C_{13}H_8)]^-$.

Set of results of Infrared-, Raman-, MASS-, NMR $-^{13}C-$ and RFES has led to a conclusion, that the organic part of a composite represents polyacetylene a trans-structure. Polymeric chains lace, and alongside with the interfaced double bonds, are present linear fragments of twinned double $-HC=C=CH-$ and triple $C\equiv C$ bonds.

Reactions of $NbCl_5$ with acetylene also is applied to synthesis of the organo-inorganic composites containing in an organic matrix clusters of transitive elements not only VI, but other groups of periodic system.

Solutions of $MoCl_5$ stated at a room temperature in benzene or toluene used for reception of organo-inorganic composites with enough high concentration of metal. Acetylene passed through these solutions during 4–6 h. Acetylene preliminary refined and drained from water and possible impurity. Color of a solution is varied from dark-yellow-green up to black in process of transmission acetylene. The solution heated up, turned to gel of black color and after a while the temperature dropped up to room. Reaction was accompanied by formation HCl. Completeness of interaction pentachloride with acetylene judged on the termination of its allocation.

The sediment, similar to gel, settled upon termination of transmission acetylene. It filtered off in an atmosphere of argon, washed out dry solvent and dried up under vacuum.

The received substances are fine-dispersed powders of black color, insoluble in water and in usual organic solvents. Solutions after branch of a deposit represented pure solvent according to NMR. Formed compounds of molybdenum and products of oligomerization acetylene precipitated completely. The structure of products differed under the maintenance of carbon depending on speed and time of transmission acetylene a little. Thus relation C:H was conserved close to unit, and Cl:Mo—close to two. The structure of products of reaction differed slightly in benzene and toluene (Table 4.1) and was close to $MoCl_{1.9 \pm 0.1}(C_{30 \pm 1}H_{30 \pm 1})$.

TABLE 4.1 Data of the element analysis of products for reaction $MoCl_5$ with acetylene in benzene and toluene

Solution	C		H		Cl		Mo	
	Findings	Calculated	Findings	Calculated	Findings	Calculated	Findings	Calculated
Benzene	65.94		5.38		11.70		16.98	
		64.67		5.38		12.73		17.22
Toluene	66.23		5.24		11.87		16.66	

Per centage, weight (%)

Presence on diffraction patterns the evolved products of a wide maximum at small corners allowed to assume X-ray amorphous or nanocrystalline a structure of the received substances. By a method of scanning electronic microscopy (SEM) it was revealed, that substances have low crystallinity and nonfibrillary morphology (Figure 4.1(a)).

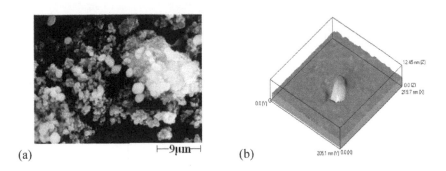

(a) |—9μm—| (b)

FIGURE 4.1   Photomicrographes $MoCl_{1.9 \pm 0.1}(C_{30 \pm 1}H_{30 \pm 1})$ according to (a) SEM and (b) ASM.

4.3 PHYSICOCHEMICAL PROPERTIES AND APPLICATION OF POLYACETYLENE

By results of atomicpower microscopy (ASM) particle size can be estimated within the limits of $10 \div 15$ nm (Figure 4.1(b)). By means of translucent electronic microscopy has been positioned, that the minimal size of morphological element $MoCl_{1.9 \pm 0.1} (C_{30 \pm 1}H_{30 \pm 1})$ makes $1 \div 2$ nm.

Substances are steady and do not fly in high vacuum and an inert atmosphere up to 300°C. Formation of structures $[Mo_{12}Cl_{24}(C_{20}H_{21})]^-$ and $[Mo_{13}Cl_{24}(C_{13}H_8)]$ is supposed also on the basis of mass-spectral of researches.

Spectrum EPR of composite $MoCl_{1.9 \pm 0.1} (C_{30 \pm 1}H_{30 \pm 1})$ at 300 K (Figure 4.2(a)) consists of two isotropic lines. The intensive line $g = 1,935$ is carried to unpaired electrons of atoms of molybdenum. The observable size of the g-factor is approximately equal to values for some compounds of trivalent molybdenum. For example, in $K_3[InCl_6] \cdot 2H_2O$, where the ion of molybdenum Mo (+3) isomorphically substitutes In (+3), and value of the g-factor makes 1.93 ± 0.06[1].

The line of insignificant intensity with $g = 2,003$, close to the g-factor-free electron 20,023, has been carried to unpaired electrons atoms of carbon of a polyacethylene matrix. Intensity of electrons signals for atoms molybdenum essentially above, than for electrons of carbon atoms of a matrix. It is possible to conclude signal strength, that the basic contribution to paramagnetic properties of a composite bring unpaired electrons of atoms of molybdenum.

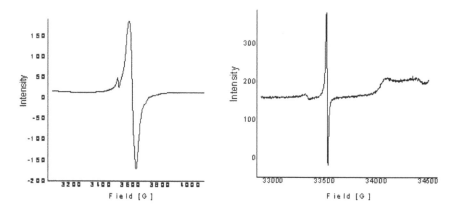

FIGURE 4.2   Spectrum EPR of composite $MoCl_{1.9 \pm 0.1}(C_{30 \pm 1}H_{30 \pm 1})$ at 300 K, removed in a continuous mode in (a)—X and (b)—W-range.

Spectrum EPR which has been removed in a continuous mode at 30 K, (Figure 4.2(b)) has a little changed at transition from X-to a high-frequency W-range. Observable three wide lines unpaired electrons atoms of molybdenum have been carried to three axial components with g_1 = 1.9528, g_2 = 1.9696 and g_3 = 2.0156 accordingly. Unpaired electrons atoms of carbon of a polyacethylene matrix the narrow signal g = 2.0033 answers. In a pulse mode of shooting of spectra EPR at 30 K (Figure 4.3) this line decomposes on two signals with g_1 = 2,033 and g_2 = 2.035. Presence of two signals EPR testifies to existence in a polyacethylene matrix of two types of the paramagnetic centers of the various nature which can be carried to distinction in their geometrical environment or to localized and delocalized unpaired electrons atoms of carbon polyacethylene chain.

Measurement of temperature dependence of a magnetic susceptibility X_g in the field of temperatures 77 ÷ 300 K has shown, that at decrease in temperature from room up to 108 K the size of a magnetic susceptibility of samples is within the limits of sensitivity of the device or practically is absent. The sample started to display a magnetic susceptibility below this temperature. The susceptibility sharply increased at the further decrease in temperature.

The magnetic susceptibility a trans-polyacetylene submits to Curie law and is very small on absolute size. Comparison of a temperature course X^χ_g

a composite and pure allows to conclude a trans-polyacetylene, that the basic contribution to a magnetic susceptibility of the investigated samples bring unpaired electrons atoms of molybdenum in cluster. Sharp increase of a magnetic susceptibility below 108 K can be connected with reduction of exchange interactions between atoms of metal.

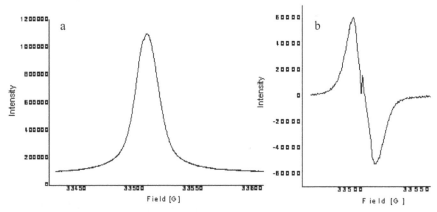

FIGURE 4.3   Spectrum absorption EPR of composite $MoCl_{1.9 \pm 0.1}(C_{30 \pm 1}H_{30 \pm 1})$, removed in a pulse mode in a W-range at (a)—30 K and (b)—its first derivative.

The size of electroconductivity compressed samples $MoCl_{1.9 \pm 0.1}(C_{30 \pm 1}H_{30 \pm 1})$, measured at a direct current at a room temperature-$(1.3 \div 3.3) \cdot 10^{-7}$ $Ohm^{-1} \cdot cm^{-1}$ is in a range of values for a trans-polyacetylene and characterizes a composite as weak dielectric or the semiconductor. The positioned size of conductivity of samples at an alternating current $\sigma = (3.1 \div 4.7) \cdot 10^{-3}$ $Ohm^{-1} \cdot cm^{-1}$ can answer presence of ionic (proton) conductivity that can be connected with presence of mobile atoms of hydrogen at structure of polymer.

Research of composition, structure and properties of products of interaction $NbCl_5$ with acetylene in a benzene solution also are first-hand close and differ a little with the maintenance of carbon (Table 4.2).

TABLE 4.2 Data of the element analysis of products of interaction $NbCl_5$ with acetylene in a solution and at direct interaction

The weights content (%)							
C		H		Cl		Nb	
Findings	Calculated	Findings	Calculated	Findings	Calculated	Findings	Calculated
In a solution							
45.00	45.20	3.60	3.76	20.50	22.20	26.50	28.80
At direct interaction							
41.60	45.20	4.02	3.76	22.81	22.20	27.78	28.80

To substances formula $NbCl_{2\pm0,1}(C_{12\pm1}H_{12\pm1})$ can be attributed on the basis of the received data. Interaction can be described by the following equation:

$$NbCl_{5(solv/solid)} + nC_2H_2 \rightarrow NbCl_{2\pm0,1}(C_{12\pm1}H_{12\pm1})\downarrow + (n-12)C_6H_6 + 3HCl\uparrow + Q$$

The wide line was observed on diffraction pattern $NbCl_{2\pm0,1}(C_{12\pm1}H_{12\pm1})$ at $2\theta = 23 - 24°C$. It allowed to assume a nanocrystalline structure of the received products.

Studying of morphology of surface $NbCl_{2\pm0,1}(C_{12\pm1}H_{12\pm1})$, received by direct interaction, method SEM has shown, that particles have predominantly the spherical form, and their sizes make less than 100 nm (Figure 4.4).

The globular form of particles and their small size testify to the big size of their specific surface. It will be coordinated with high catalytic activity $NbCl_{2\pm0.1}(C_{12\pm1}H_{12\pm1})$.

Fibrils are formed in many cases as a result of synthesis [24]. The morphology of polyacetylene films practically does not depend on conditions of synthesis. Diameter of fibrils can change depending on these conditions and typically makes 200–800 Å [14, 24]. At cultivation of films on substrates, the size fibrils decreases. The same effect is observed at reception of polyacetylene in the medium of other polymers. Time of endurance (ageing) of the catalyst especially strongly influences of the size of fibrils. The size of fibrils increases with increase in time of ageing. Detailed research of growth fibrils on thin films a method of translucent electronic microscopy has allowed to find out microfibrillar branchings on the basic fibril (the size 30–50 Å) and thickenings in places of its gearing, and also presence of rings on the ends of fibrils.

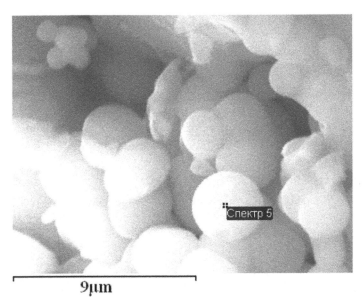

FIGURE 4.4   Microphoto of particles $NbCl_{2 \pm 0.1}(C_{12 \pm 1}H_{12 \pm 1})$, received by method SEM.

Essential changes in morphology of a film at isomerization of poly-acetylene are not observed. The film consists from any way located fibrils. Fibrils sometimes are going in larger formations [23]. Formation of a film is consequence of interaction fibrils among themselves due to adhesive forces. The big practical interest is represented catalytic system $AlR_3–Ti(oBu)_4$ [48]. Polyacetylene is received on it at −60°C, possesses fibrous structure and can be manufactured usual, accepted in technology of polymers by methods.

Particle size increases from 100 up to 500 Å at use of the mixed catalyst and the increase in density of films is observed. A filtration of suspension it is possible to receive films of any sizes. On the various substrates possessing good adhesive properties, it is possible to receive films dispersion of gel. Polymer easily doping AsF_5, $FeCl_3, I_2$, and other electron acceptors. Preliminary tests have shown some advantages of such materials at their use in accumulators [48]. Gels polyacetylene with the similar properties, received on others catalytic systems, represent the big practical interest [49]. Data on technological receptions of manufacturing of polyacetylene films from gels with diameter of particles 0.01–1.00 mm are in the patent literature [50].

Films are received at presence of the mixed catalyst at an interval of temperatures from—100 up to –48°C. These results testify to an opportunity of transition to enough simple and cheap technology of continuous process of reception of polyacetylene films. Rather accessible catalysts are WCl_6 and $MoCl_6$. Acetylene polymerizes at 20°C and pressure up to 14 atm at their presence. However the received polymers contain carbonyl groups because of presence of oxygen and possess low molecular weight [51]. The complex systems including in addition to WCl_6 or $MoCl_6$ tetraphenyltin are more perspective for reception of film materials. The film of doped predominantly (90%) a trans-structure with fibrous morphology is formed on a surface of a concentrated solution of the vanadic catalyst. Diameter of fibrils of the polymer received on catalyst $MoCl_5$-Ph_4Sn, can vary within the limits of 300–10,000 Å; in case of catalyst WCl_6-Ph_4Sn, it reaches $1,2 \cdot 10^5$ Å [30].

Research of a kinetics of polymerization of acetylene on catalysts $Ti(OBu)_5$–$Al(Et)_3$, WCl_6–PhtSn, $MoCl_5$–Ph_4Sn, $Ti(CH_2CeH_5)$, has shown, that speed of process falls in the specified number [31]. Films, doped by various acceptors $[CH (SbF_5)_{0.7}]_n$, $[CH (CF_3SO_3H)_{0.8}]_m$– had conductivity 10–20 $Ohm^{-1} \cdot cm^{-1}$ at 20°C. New original method of reception of films doped in a trans-form is polymerization of a 7,8-bis (trifluoromethyl) tricycle [4,2,2,0] deca-3,7,9-trien (BTFM) with disclosing a cycle.

Polymerization occurs on catalytic system WCl_6-$Sn(CH_3)_4$, precipitated on surfaces of a reactor. The film prepolymer as a result of heating in vacuum at 100–150° detaches trifluoromethyl-benzene. The silvery film polyacetylene is formed. The density of polymer reaches 1, 1 g/cm^3 and comes nearer to flotation density polyacetylene, received in other ways. Received this method of amorphous polyacetylene has completely a trans-configuration and does not possess fibrous structure. The rests of 1,2-bis(trifluoromethyl-benzene) are present at polymer according to infrared-spectroscopy. Crystal films of polyacetylene with monoclinic system and $\beta = 91,5°$ are received at long heating of prepolymer on networks of an electronic microscope at 100° in vacuum. Improvement of a method has allowed to receive films and completely oriented crystal polyacetylene. In the further ways of synthesis of polymers from others monomers have been developed.

Naphthalene evolves at heat treatment of prepolymer in the first case. Anthracene evolves in the second case. Purification polymer from residual impurities occurs when the temperature of heat treatment rise. According to spectral researches, the absorption caused by presence of sp-hybrid carbon is not observed in films. Absorption in the field of 1,480 cm^{-1} caused by presence of C=C bonds in Spectra KP of considered polymers, is shifted compared to the absorption observed in polyacetylene, received by other methods (1,460 cm^{-1}).

It is believed that it is connected with decrease of size of interface blocks. The obtained films are difficult doping in a gas phase due to its high density. Conductivity of initial films reaches 10–200 (Ohm·cm^{-1}) when doped with bromine or iodine in a solution [51].

Solutions of complex compound cyclopentadienyl-dititana in hexane and sodium cyclopentadienyl complex have high catalytic activity. Films with a metallic luster can be obtained by slowly removing the solvent from the formed gel polyacetylene in vacuum. It is assumed that the active complex has a tetrahedral structure. Polymerization mechanism is similar to the mechanism of olefin polymerization on catalyst Ziegler–Natta. Obtained at −80°C cis-polyacetylene films after doping had a conductivity of 240 (Ohm·cm^{-1}).

Classical methods of ionic and radical polymerization do not allow to receive high-molecular polymers with system of the conjugated bonds because of isomerization the active centers [2] connected with a polyene chain. Affinity to electron and potential of ionization considerably varies with increase in effective conjugated. One of the methods, allowing to lead a cation process of polymerization, formation of a complex with the growing polyconjugated chains during synthesis. Practical realization of such process probably at presence of the big surplus of a strong acceptor of electrons in the reaction medium. In this case, the electronic density of polyene chain falls. The probability of electron transfer from a chain on the active center decreases accordingly.

Polyacetylene films were able to synthesize on an internal surface of a reactor in an interval of temperatures from −78 up to −198°C at addition of acetylene to arsenic pentafluoride. Strips of absorption cis-polyacetylene are identified in the field of 740 cm^{-1} and doped complexes in the field of 900 and 1,370 cm^{-1}.

Similar in composition films were prepared by polymerization vinylacetylene in the gas phase in the presence SbF_5. However, to achieve a metallic state has failed. Soluble polyenes were obtained in solution AsF_3 in the by cationic polymerization, including soluble agents. Polymerization was carried out at the freezing temperature of acetylene. From the resulting solution were cast films with low conductivity, characteristic of weakly doped polyacetylene with 103 molecular weight. Spectral studies confirmed the presence of the polymer obtained in the cis-structure. Practical interest are insoluble polymers with a conductivity of 10^{-3} (Ohm-cm^{-1}), obtained at $-78°C$ polymerization of acetylene in the presence AsF_5, Na-AsF, SiF, AsF_3, BF_3, SbF_5, and PF_6.

Effective cocatalyst of cationic polymerization of acetylene and its derivatives are compounds of bivalent mercury and its organic derivatives. As a result, the reactions of oxide or mercury salts with proton and aprotic acids into saturated hydrocarbon formed heterogeneous complexes. They are effectively polymerized acetylene vinylacetylene, phenylacetylene, propargyl alcohol. Polyacetylene predominantly trans-structure with a crystallinity of 70 per cent was obtained in the form of films during the polymerization of acetylene on the catalyst surface. According to X-ray studies, the main reflection corresponds to the interplanar distance d=3.22 Å. Homogeneous catalyst obtained in the presence of aromatic ligands. Active complex of the catalyst with acetylene is stable at low temperatures. The alkylation of solvent and its interpolymerization with acetylene is in the presence of aromatic solvents (toluene, benzene). In the infrared spectrum of the films revealed absorption strias corresponding to the aromatic cycle. Patterns of polymerization of acetylene monomers, the effect of temperature and composition of the catalyst on the structure and properties of the resulting polymers were studied. Morphology of the films showed the absence of fibrils.

Almost all of the above methods of synthesis polyacetylene with high molecular weight lead to the formation of insoluble polymers. Their insolubility due to the high intermolecular interaction and form a network structures. For the polymers obtained by Ziegler catalyst systems, the cross-link density of the NMR data of 3–5 per cent. This is confirmed by ozonolysis. Quantum chemical calculations confirm that the isomerization process intermolecular bonds are formed.

The most accessible and promising method for synthesis of the polyacetylene of linear structure is the polymerization of acetylene in the presence of metals VIII group, in the presence with reducing agents (catalyst Luttlnger). Polyacetylene with high crystallinity was obtained by polymerization of acetylene on the catalyst system $Co(NO_3)_2$–$NaBH_4$ with component ratio 1:2. Both components are injected into the substrate containing monomer, to prevent the death of a catalyst. Raising the temperature and the concentration of sodium borane leads to partial reduction of the polymer. The activity of nickel complex can be significantly improved if the polymerization leads in the presence of $NaBH_2$. Crystalline polymers, obtained at low temperatures, do not contain fibrils. Crystallites have dimensions of 70 Å.

A typical reflex observed at 23.75°C (d=3.74 Å) confirms the trans-structure of the polymer. Palladium complexes are ineffective in obtaining high molecular weight polymers. Catalyst Luttlnger enables one to obtain linear polymers of cis-structure, characterized by high crystallinity. The yield of polyacetylene is 25–30 g/g catalyst.

Chlorination of the freshly prepared polymers at low temperatures allows to obtain soluble chlorpolymers about 104 molecular weight [27]. Although the authors argue that the low-temperature chlorination, in contrast to hydrogenation, there is no polymer degradation, data suggest that an appropriate choice of temperature and solvent derived chlorinated polymers have a molecular weight up to $2,5\cdot10^5$. Destruction more visible at chlorination on light and on elevated temperatures. The proof of the linear structure of polyacetylene is the fact that soluble iodinated polymers are obtained by iodination polyacetylene suspension in ethanol.

Systematic studies of methods for the synthesis of polyacetylene allowed to develop a simple and convenient method of obtaining the films on various substrates wetted by an ethereal solution of the catalyst. The disadvantages of these films, as well as films produced by Shirakava [24, 25] are difficult to clean them of residual catalyst and the dependence of properties on the film thickness. Much more manufacturable methods for obtaining films of polyacetylene spray pre-cleaned from residues of the catalyst in a stream of polyacetylene gels inert gas or a splash of homogenized gels [49]. The properties of such films depend on the conditions of their formation. Free film thickness of 2–3 mm are filtered under pressure in an inert atmosphere containing a homogenized suspension of polyacetylene 5–10 g/liter. The suspension formed in organic media at low temperatures in the presence of the catalyst $Co(NO_3)_2$-$NaBH_4$. The films obtained by spraying a stream of inert gas, homogeneous, have good adhesion to substrates made of metal, polyurethane, polyester polyethyleneterephtalate, polyimide, etc.

Suspension of polyacetylene changes its properties with time significantly. Cross-linking and aggregation of fibrils observed in an inert atmosphere at temperatures above −20°C. This leads to a decrease in the rate of oxidation and chlorination. The morphology of the films changes particularly striking during the aging of the suspension in the presence of moisture and oxygen: increasing the diameter of the fibrils, decreasing their length, breaks and knots are formed. The suspension does not change its properties in 2 weeks.

Preparation of soluble polymers with a system of conjugated double bonds, and high molecular weight is practically difficult because of strong intermolecular interactions. Sufficiently high molecular weight polyacetylene were obtained in the form of fine-dispersed particles during the synthesis on a Luttinger catalyst in the presence surfactants: copolymer of styrene and polyethylene with polyethylene oxide. Acetylene was added to the catalyst and the copolymer solution in a mixture of cyclohexane-tetrahydrofuran at −60°C and heated to −30°C. Stable colloidal solutions with spherical particles in size from 40 to 2,000 Å, formed. The density of selected films is 1.15 g/cm^3. Colloidal solutions of polyacetylene can be obtained in the presence of other polymers that prevent aggregation of the forming molecules of polyacetylene.

Polyacetylene obtained in the presence of Group VIII metals, in combination with NaBH$_4$, has almost the same morphology, as a polymer synthesized by Shirakawa [25]. The dimensions of fibrils lay in the range 300–800 Å, and depend on the concentration of the catalyst, the synthesis temperature and medium [27].

Thermogravimetric curves for the polyacetylene, there are two exothermic peaks at 145 and 325°C [16]. The first of these corresponds to an irreversible cis-trans isomerization. Migration of hydrogen occurs at 325°C, open chain and cross-linking without the formation of polyacetylene volatile products. The color of the polymer becomes brown. A large number of defects appears. In the infrared spectrum there are absorption bands characteristic of the CH$_2$, CH$_3$, -C=C- and -C$_2$H$_5$- groups [16].

Structuring polymer occurs in the temperature range 280–380°. But 72 per cent of initial weight of polyacetylene losses at 720°C. The main products of the decomposition of polyacetylene are benzene, hydrogen and lower hydrocarbons [15]. The crystallinity of polyacetylene reduced when heated in air to 90° after several hours. The brown amorphous substance, similar cuprene, obtained after 70 hours.

Catalytic hydrogenation of polyacetylene leads to the formation of cross-linked product [29]. Non-crosslinked and soluble products are obtained in the case of hydrogenation of polyacetylene doped with alkali metals [30, 31]. Polyacetylenes are involved in redox reactions that occur in processing strong oxidizing and reducing agents (iodine, bromine, AsF_5, Na-naphthalene in order to significantly increase the electrical conductivity [32].

Practical use of polyacetylene is complicated by its easy oxidation by air oxygen [33]. Oxidation is easily exposed to the polymer obtained by polymerization of acetylene [34, 35]. The cis-or trans $-(CH)_x$ in air or oxygen for about an hour exposed to the irreversible oxidative degradation [36, 37]. The limiting value of weight gain due to absorption of polyacetylene (absorption) of oxygen from air oxidation at room temperature is 35 per cent [33]. The resulting product is characterized by the formula $[(C_2H_2)O_{0,9}]_n$. The ease of oxidation depends on the morphology of the polymer and changes in the series of crystal amorphous component the surface of the fibrils [35]. The absorption of oxygen begins at the surface of fibrils, and then penetrates. Polyacetylene globular morphology is more stable to the effects of O_2 than polymer fibrillar structures [38].

Polyacetylene obtained by polymerization of acetylene in Ziegler–Natta catalysts, after doping Cl_2, Br_2, I_2, AsF_5 becomes a semiconductor in the form of flexible, silvery films "organic metals" [39]. Doping with iodine increases the amorphous samples δ $6·10^{-5}$, and crystal to $7·10^2$ $Ohm^{-1}·cm^{-1}$ [40]. The highest electrical conductivity of the polyacetylene compared with those obtained by other methods, the authors [41] explain the presence of catalyst residues. In their view, the concentration of the structure of sp^3-hybridized carbon atoms is relatively little effect on the conductivity as compared with the influence of catalyst residues. Doping with iodine films of polyacetylene obtained by metathesis polymerizing cyclooctatetraene leads to an increase in their electrical conductivity 10^{-8} to $50–350 \ Ohm^{-1}·cm^{-1}$ [42], and have received polymerization of benzvalene with ring opening from $10^{-8}–10^{-5}$ to $10^{-4}–10^{-1} \ Ohm^{-1}·cm^{-1}$ [43].

Conductivity increases when pressure is applied to polyacetylene, obtained by polymerization of acetylene [44] and interphase dehydrochlorination of PVC [45]. Anomalously large (up to ten orders of magnitude) an abrupt increase in conductivity when the load is found for iodine-doped crystalline polivinilena—conversion product of PVC [46].

Magnetic properties of polyacetylene significantly depend on the configuration of chains [47]. In the EPR spectrum of the polymerization of polyacetylene singlet line with g-factor of 2.003 [48] and a line width (ΔH) of 7 to 9.5 Oe for the cis-isomer [49] and from 0.28 to 5 Oe for the trans-isomer [50] observed. According to other reports [47], the cis-isomer, syn-synthesized by polymerization of acetylene at 195 K, the EPR signal with g-factor = 2.0025 is not observed. This signal appears when the temperature of polymerization increases, when the trans-isomer in the form of short chains mainly at the ends of the molecules is 5–10 wt per cent [51]. The morphology of polyacetylene also has an effect on the paramagnetic properties. The concentration of PMC in the amorphous polyacetylene is ~1,018 spin/g, and in the crystal −1,019 spin/g [51].

KEYWORDS

- *Carboneum*
- Carbyne
- Physicochemical properties
- Polyacetylene

REFERENCES

1. Sladkov, A. M.; Kasatochkin, V. I.; Korshak, V. V.; and Kudryavcev Yu. P.; Diploma on discovery no 107. *Bull. Invent.* **1972**, 6.
2. Korshak, V. V.; Kasatochkin, V.I.; Sladkov, A. M.; Kudryavcev, Yu. P.; and Usunbaev, K.; About Synthesis and Properties of Polyacetylene. Lecture Academy of Sciences the USSR; **1961**, *136(6)*, 1342.
3. Sladkov, A. M.; Carbyne— the Third Allotropic form of Carbon. M.: Science; **2003**, 152 p.
4. Heimann, R. B.; and Evsyukov, S. E.; Allotropy of carbon. *Nature.* **2003**, *8*, 66.
5. Sladkov, A. M.; and Kudryavcev Yu. P; Diamond, graphite, carbyne-allotropic forms of carbon. *Nature.* **1969**, *5*, 37.
6. Kudryavcev, Yu. P.; Evsyukov, S. E.; Guseva, M. B.; Babaev, V. G.; and Xvostov, V. V.; Carbyne— the Third Allotropic form of Carbon. Proceedings of the Academy of Sciences, Series Chemical; **1993**, *3*, 450 p.

7. Natta, G.; Pino, P.; and Mazzanti, G.; Patent. Hal. 530753 Italy C. A.; **1958**. *52*. 15128 p.
8. Natta, G.; Mazzanti, G.; and Corradini, P.; *AIII Accad. Naz. Lincei. Cl. Sci. Fis. Mat. Nat. Rend.* **1958,** *25*, 2.
9. Watson, W. H.; Memodic, W. C; and Lands, L. G.; *3 Polym. Sci.* **1961,** *55*, 137.
10. Shirakawa, H.; and Ikeda, S.; *Polym. J.* **1971,** *2*, 231.
11. Ito, T.; Shirakawa, H.; and Ikeda, S. J.; *Polym. Sci. Polym. Chem. Ed.* **1974,** *12*, 11.
12. Tripathy, S. K.; Rubner, M.; and Emma, T.; et al. Ibid. **1983,** *44*, C3—37.
13. Wegner, G.; *Macromol. Chem.* **1981,** *4*, 155.
14. Schen, M. A.; Karasz, F. E.; and Chien, L C.; *J. Polym. Sci.: Polym. Chem. Ed.* **1983,** *21*, 2787.
15. Wnek, G. E.; Chien, J. C; Karasz, F. E.; et al. *J. Polym. Sci. Polym. Lett. Ed.* **1979**, *17*, 779.
16. Aldissi, M.; Synthetic Metals. **1984,** *9*, 131.
17. Schue, F.; Aldissi, A. F.; Colloq. Int. Nouv. Orient. Compos. Passifs. Mater. Technol. Mises Ocure. Paris; **1982,** 225 p.
18. Chien, M. A.; Karasz, F. E.; and Chien, J. C.; *Macromol. Chem. Rapid Commun.* **1984,** *5*, 217.
19. Chien, J. C.; *J. Poli. Sci. Polym. Lett. Ed.* **1983,** *21*, 93.
20. Saxman, A. M.; Liepins, R.; Aldissi, M.; *Progr. Polym. Sci.* **1985,** *11*, 57.
21. Chien, J. C.; Karasz, F. E.; Schen, M. A.; and Hirsch, T.; 4. *Macromole.* **1983,** *16*, 1694.
22. Chien, J. C.; Karasz, F. E.; MacDiarmid, A. G.; and Heeger, A. J.; *Polym. Sci. Polym. Lett. Ed.* **1980,** *18*, 45.
23. Chien, J. C.; Polym. News. **1979,** *6*, 53.
24. Dandreaux, G. F.; Galuin, M. E.; Wnek, G. E.; *J. Phys.* **1983,** *44*, C3—135.
25. Parshakov, A. S.; Abstract of a Thesis Interaction Pentachloride Molybdenum with Acetylene a New Method for the Synthesis of Nanoscale Composite Materials. Moscow: IONCh RAS; **2010**. Ilin, E. G.; Parshakov, A. S.; Buryak, A.K.; Kochubei, D.I.; Drobot, D. V.; Nefedov Dan, V. I.; **2009,** *427(5)*, 641–645 p.
26. Matnishyan, A. A.; *Adv. Chem.* **1988,** *57(4)*, 656–683.
27. Natta, G.; Pino, P.; and Mazzanti, G.; Patent. Hal. 530753 Italy, C. A.; **1958,** *52*, 15128 p.
28. Natta, G.; Mazzanti, G.; and Corradini, P.; *AIII Accad. Naz. Lincei, Cl. Sci. Fis. Mat. Nat. Rend.* **1958,** *25*, 2.
29. Chasko, B.; Chien, J. C. W.; Karasz, F. E.; Mc Diarmid, A. G.; and Heeger, A. J.; *Bull. Am. Phys. Soc.* **1979,** *24*, 480–483
30. Shirakawa, H.; Sato, M.; Hamono, A.; Kawakami, S.; Soga, K.; and Ikeda, S.; *Macromolec.* **1980,** *13(2)*, 457–459.
31. Soga, K.; Kawakami, S.; Shirakawa, H.; and Ikeda, S.; *Makromol. Chem. Rapid. Commun.* **1980,** *1(10)*, 643–646.
32. Lopurev, V. A.; Myachina, G. F.; Shevaleevskiy, O. I.; and Hidekel, M. L.; *High-Mole. Compounds, A.* **1988,** *30(10)*, 2019–2037.
33. Kobryanskiy, V. M.; Zurabyan, N. J.; Skachkova, V. K.; and Matnishyan, A. A.; High-Molecular Compounds, B. **1985,** *27(7)*, 503–505.

34. Berlyn, A. A.; Geyderih, M. A.; and Davudov, B. E.; Chem. Polyconjugate Systems. M.: Chem; **1972**, 272 p.

35. Yang, X. Z.; and Chien, J. C. W.; *J. Polym. Sci.: Polym. Chem. Ed.* **1985**, *23(3),* 859–878.

36. Mc Diarmid, A. G.; Chiang, J. C.; and Halpern, M.; et al. *Am. Chem. Soc. Polym. Prepr.* **1984**, *25(2),* 248–249.

37. Gibson, H.; and Pochan, J.; *Macromole.* **1982**, *15(2),* 242–247.

38. Kobryanskii, V. M.; *Mater. Sci.* **1991**, *27(1),* 21–24.

39. Deits, W.; Cukor, P.; Rubner, M.; and Jopson, H.; *Electron. Mater.* **1981**, *10(4),* 683–702.

40. Heeger, A. J.; Mc Diarmid, A. G.; and Moran, M. J.; *Am. Chem. Soc. Polym. Prepr.* **1978**, *19(2),* 862.

41. Arbuckle, G. A.; Buechelev, N. M.; and Valentine, K. G.; *Chem. Mater.* **1994**, *6(5),* 569–572

42 Korshak, J. V.; Korschak, V. V.; Kanischka, Gerd Hocker Hartwig Makromol. *Chem. Rapid Commun.* **1985**, *6(10),* 685–692.

43. Swager, T. M.; and Grubbs, R. H.; *Synth. Met.* **1989**, *28(3),* D57–D62, 51.

44. Matsushita, A.; Akagi, K.; Liang, T. S.; and Shirakawa, H.; *Synth. Met.* **1999**, *101(1–3),* 447–448.

45. Salimgareeva, V. N.; Prochuhan, Yu. A.; and Sannikova, N. S.; *High-mole. Compounds.* **1999**, *41(4),* 667–672.

46. Leplyanin, G. V.; Kolosnicin, V. S.; and Gavrilova, A. A.; et al. *Electro Chem.* **1989**, *25(10),* 1411–1412.

47. Zhuravleva, T. S.; *Adv. Chem.* **1987**, *56(1),* 128–147.

48. Goldberg, I. B.; Crowe, H. R.; Newman, P. R.; Heeger, A. J.; and Mc Diarmid, A. G. J.; *Chem. Phys.* **1979**, *70(3),* 1132–1136.

49. Bernier, P.; et al. *Polym. J.* **1981,** *13(3),* 201–207.

50. Holczer, K.; Boucher, J. R.; Defreux, F.; and Nechtschein, M.; *Chem. Scirpta.* **1981,** *17(1–5),* 169–170.

51. Krinichnui, V. I.; *Adv. Chem.* **1996**, *65(1),* 84.

CHAPTER 5

TRENDS IN ACTIVATED CARBON FIBERS

M. MEHDIPOUR

CONTENTS

5.1   INTRODUCTION

In this chapter, we focus on the pore size controlling in carbon-based nanoadsorbent to apply simulation and modeling methods and describe the recent activities about it. Significant progress has been made in this process throughout recent years. Essential emphasis is put on the controlling of both micro and mesoporosity and its applications. For the control of mesopores, many novel methods are proposed such as catalytic activation, polymer blend, organic gel, and template carbonization, for analyzing micropore distribution in activated carbons assumes an array of semi-infinite, rigid slits of distributed width whose walls are modeled as energetically uniform graphite. Various kinds of pores in solid materials are classified into intraparticle pores and interparticle pores according to the origin of the pores and the structural factors of the pores are discussed as well as the methods for evaluation of the pore-size distribution with molecular adsorption (molecular resolution porosimetry), small angle X-ray scattering, mercury porosimetry, nuclear magnetic resonance, and thermoporosirmetry [1–46]. The main aim in the controlling of micropores is to produce molecular sieving carbon (MSC) with uniform micropore structure, that applied in special membranes or produce some carbon composites that remove special contaminate from aqueous environments as adsorbent [47–53]. Despite difference in particle size, the adsorption properties of activated carbon and carbon based nanoadsorbents(ACF–ACNF–CNF and CNT), are basically the same because the characteristics of activated carbon (pore size distribution, internal surface area,and surface chemistry) controlling the equilibrium aspects of adsorption are independent of particle [54–120]. The excellently regular structures of CNTs and other carbon nanostructures and its composites facilitate accurate simulation of CNTs' behavior by applying a variety of mathematical, classical, and numerical methods and simulation, Such as Grand Canonical Monte Carlo (GCMC) simulation, Car—Parrinello molecular dynamics (CPMD), the ab initio density functional theory (DFT), atomistic and molecular dynamics simulation (MDS), LJ potential, HK Method, BJH Method, DR, etc. Adsorption isotherms can be simulated and modeled for this system using GCMC or DFT and other methods, pore-size distributions and PSD curves are determined from experimental isotherms and using such models and finally experimental and model results of carbon material samples is compared, adapted and model is verified [121–145].

5.1.1 ACTIVATED CARBON: PROPERTIES AND APPLICATION

TABLE 5.1 Technical advantages and disadvantages of existing modification techniques

Modification	Treatment	Advantages	Disadvantages
Chemical characteristics	Acidic	Increases acidic functional groups on ac surface. Enhances chelation ability with metal species	May decrease bet surface area and pore volume, may give off undesired so_2 (treatment with h_2so_4) or no_2 (treatment with hno_3) gases, Has adverse effect on uptake of organics
	Basic	Enhances uptake of organics	May, in some cases, decrease the uptake of metal ions
	Impregnation of foreign material	Enhances in-built catalytic oxidation capability	May decrease BET surface area and pore volume
Physical characteristics	Heat	Increases BET surface area and pore volume	Decreases oxygen surface functional groups
Biological characteristics	Bioadsorption	Prolongs ac bed life by rapid oxidation of organics by bacteria before the material can occupy adsorption sites	Thick biofilm encapsulating ac may impede diffusion of adsorbate species

Activated carbon (AC) has been most effective adsorbent for the removal of a wide range of contaminants from aqueous or gaseous environment. It is a widely used adsorbent in the treatment of wastewaters due to its exceptionally high surface areas which range from 500 to 1,500 m^2g^{-1}, well-developed internal microporosity structure [1]. While the effectiveness of ACs to act as adsorbents for a wide range of pollutant materials is well noted and more research on AC modification are presented due to the need to enable ACs to develop affinity for special contaminants removal from wastewater [2]. It is, therefore, essential to understand the various important factors that influence the adsorption capacity of AC due to their modification so that it can be tailored to their specific physical and chemical attributes to enhance their affinities pollutant materials. These factors include specific surface area, pore-size distribution, pore volume, and presence of surface functional groups. In general, the adsorption capacity increases with specific surface area due to the availability of adsorption site, whereas pore size and micropore distribution are closely related to the composition of the AC, the type of raw material used, and the degree of activation during production stage [3]. Here, we summarize the various AC modification techniques and their effects on adsorption of chemical species from aqueous solutions, modification of AC in granular or powdered form were reviewed. Based on extensive literature reviews, the authors have categorized the techniques into three broad groups, namely, modification of chemical, physical, and biological characteristics that are further subdivided into their pertinent treatment techniques (Table 5.1 lists and compares the advantages and disadvantages of existing modification techniques with regards to technical aspects that are further elucidated in the following sections [1–4].

While these characteristics are reviewed separately as reflected by numerous AC modification research, it should be noted that there were also research with the direct intention of significantly modifying two or more characteristic and that the techniques reviewed are not intended to be exhaustive. The adsorption capacity depends on the accessibility of the organic molecules to the microporosity that is dependent on their size [5]. AC can be used for removing taste and odor (T&O) compounds, synthetic organic chemicals (SOCs), and dissolved natural organic matter (DOM) from water.

PAC typically has a diameter less than 0.15 mm, and can be applied at various locations in a treatment system. GAC, with diameters ranging from 0.5 to 2.5 mm, is employed in fixed-bed adsorbents such as granular media filters or post filters. Each of these factors must be properly evaluated in determining the use of activated carbon in a practical application. The primary treatment objective of activated carbon adsorption in a particular water treatment plant determines the process design and operation; multiple objectives can not, in most cases, be simultaneously optimized [5–40].

It is critical in either case to understand and evaluate the adsorption interactions in the context of drinking water treatment systems. Activated carbons are prepared from different precursors and used in a wide range of industries. Their preparation, structure, and applications were reviewed in different books and reviews. High BET surface area and light weight are the main advantages of activated carbons. Usually activated carbons have a wide range of pore sizes from micropores to macropores, which shows a marked contrast to the definite pore size of zeolites [41–46].

5.2 ACTIVATED CARBON FIBERS: PROPERTIES AND APPLICATION

Activated carbons fibers (ACFs) have been prepared recently and developed a new field of applications. They have a number of advantages over granular activated carbons. The principal merit to prepare activated carbon in fibrous morphology is its particular pore structure and a large physical surface area. Granular activated carbons have different sizes of pores (macropores, mesopores, and micropores), whereas ACFs have mostly micropores on their surfaces. In granular activated carbons, adsorbates always have to reach micropores by passing through macropores and mesopores; whereas in ACFs, they can directly reach most micropores because micropores are open to the outer surface and hence, exposed directly to adsorbates. Therefore, the adsorption rate, as well as the amount of adsorption, of gases into ACFs is much higher than those into granular activated carbons [1–3]. In recent work, the amount of adsorption of toluene molecules is much higher, and desorption proceeds faster in ACFs than granular activated carbons, effective elimination of SO from exhausted gases by using ACFs was too. A very high specific surface area up to 2,500 m^2g^{-1} and a high micropore volume up to 1.6 cm^3g^{-1} can be obtained in isotropic-pitch-based carbon fibers. For the preparation of these carbon fibers with a very high surface area such as 2,500 m^2g^{-1}, precursors that give a carbon with poor crystallinity are recommended; thus, mesophase-pitch-based carbon fibers did not give a high surface area, whereas isotropic pitch-based carbon fibers did.

Other advantage of ACFs is the possibility to prepare woven clothes and nonwoven mats, which developed new applications in small purification systems for water treatment and also as a deodorant in refrigerators in houses, recently reported. In order to give the fibers an antibacterial function and to increase their deodorant function, some trials on supporting minute particles of different metals, such as Ag, Cu and Mn, were performed. Table 5.2 presented comparison between properties of activated carbon fibers and granular activated carbons [13–46].

TABLE 5.2 Comparison between properties of activated carbon fibers and granular activated carbons

	Activated carbon fibers	Granular activated carbons
	700–2,500	900–1,200
Surface area	0.2–2.0	~0.001
mean diameter of pores (nm)	<40	from micro- to macropores

5.3 MOLECULAR SIEVING CARBONS: PROPERTIES AND APPLICATION

Molecular sieving carbons (MSCs) have a smaller pore size with a sharper distribution in the range of micropores in comparison with other activated carbons for gas and liquid-phase adsorbates. They have been used for adsorbing and eliminating pollutant samples with a very low concentration (ethylene gas adsorption to keep fruits and vegetables fresh, filtering of hazardous gases in power plants, etc.) An important application of these MSCs was developed in gas separation systems [1–2].

The adsorption rate of gas molecules, such as nitrogen, oxygen, hydrogen, and ethylene, depends strongly on the pore size of the MSC; the adsorption rate of a gas becomes slower for the MSC with the smaller pore size. The temperature also governs the rate of adsorption of a gas because of activated diffusion of adsorbate molecules in micropores: the higher the temperature, the faster the adsorption [47–49].

By controlling (swinging) these parameters, temperature and pressure of adsorbate gas, gas separation can be performed. Depending on which parameter is controlled, swing adsorption method is classified into two modes: temperature swing adsorption (TSA) and pressure swing adsorption (PSA). Adsorption of oxygen into the MSC completes within 5 min, but nitrogen is adsorbed very slowly, less than 10 per cent of equilibrium adsorption even after 15 min. From the column of MSC, therefore, nitrogen rich gas comes out on the adsorption process, and oxygen-rich gas is obtained on the desorption process. By using more than two columns of MSC and repeating these adsorption/desorption processes, nitrogen gas is isolated from oxygen. This swing adsorption method for gas separation has advantages such as low-energy cost, room temperature operation, and compact equipment [50–53].

5.4   IMPORTANCE OF CONTROLLING OF PORE SIZE IN POROUS CARBON MATERIALS AND SUPERCAPACITORS

The main objective of this chapter is to provide a brief review of the pore-size control that is an important factor, influencing application of activated carbon or carbon nanostructures adsorption in drinking water treatment or other adsorbent application. Different pore sizes in carbon materials are required for their applications. Therefore, the PSD in carbon materials has to be controlled during their preparation, by selecting the precursor, process, and condition of carbonization, and also those of activation. A wide-range broad distribution in pore size and shape is usually obtained in carbon materials. The control of pore size in carbons is essential to compete in adsorption performance with porous inorganic materials such as silica gels and zeolites, and to use the advantages of carbon materials such as high chemical stability, high temperature resistance, and low weight.

For applications in modern technology fields, not only high surface area and large pore volume, but also a sharp pore-size distribution at a definite size and control of surface nature of pore walls are strongly required. In order to control the pore structure in carbon materials, studies on the selection of precursors and preparation conditions have been extensively carried out and certain successes have been achieved [1–3]. Pore sizes and their distributions in adsorbents have to comply with requirements from different applications. Thus, relatively small pores are needed for gas adsorption and relatively large pores for liquid adsorption, and a very narrow PSD is required for molecular sieving applications. Macropores in carbon materials were found to be effective for sorption of viscous heavy oils. Recent novel techniques to control pore structure in carbon materials can be expected to contribute to overcome this limitation [41–46].

One of newest applications that shows importance of pore structure control in carbon materials is an electric double-layer capacitor (EDLC) or supercapacitor that is an energy storage device that utilizes the EDL formed at the interface between an electrode and the electrolyte. EDLCs are well documented to exhibit significantly higher specific powers and longer cycle lifetimes compared with those of most of rechargeable batteries, including lead acid, Ni-MH, and Li-ion batteries. Hence, EDLCs have attracted considerable interest, given the ever increasing demands of electric vehicles, portable electronic devices, and power sources for memory backup.

The capacitance of an EDLC depends on the surface area of the electrode materials. Therefore, activated carbons are necessary materials for EDLC electrodes because of their large surface area, highly porous structure, good adsorption properties, and high electrical conductivity. The electrochemical performance of EDLCs is related to the surface area, the pore structure, and the surface chemistry of the porous carbon. Various types of porous carbon have been widely studied for use as electrode materials for EDLCs.

Their unusual structural and electronic properties make the carbon nanostructures applicable in, inter alia, the electrode materials of EDLCs and batteries. Activated carbon nanofibers are expected to be more useful than spherical activated carbon in allowing the relationship between pore structure and electrochemical properties to be investigated to prepare the polarizable electrodes for experimental EDLCs, EDLCs are well documented to exhibit significantly higher specific powers and longer cycle lifetimes compared with those of most of rechargeable batteries, including lead acid, Ni-MH, and Li-ion batteries [20–34–45].

Hence, EDLCs have attracted considerable interest, given the ever increasing demands of electric vehicles, portable electronic devices, and power sources for memory backup. The capacitance of an EDLC depends on the surface area of the electrode materials. Therefore, activated carbons are necessary materials for EDLC electrodes because of their large surface area, highly porous structure, good adsorption properties, and high electrical conductivity. The electrochemical performance of EDLCs is related to the surface area, the pore structure, and the surface chemistry of the porous carbon.

Various types of porous carbon have been widely studied for use as electrode materials for EDLCs. Their unusual structural and electronic properties make the carbon nanostructures applicable in the electrode materials of EDLCs and batteries. The principle of electrochemical capacitors, physical adsorption/desorption of electrolyte ions in solution, was applied for water purification by using different carbon materials [108–113].

This work is concerned with such pore control methods proposed by researchers that their ultimate goal is to establish a method with tailoring carbon material pore structures to reach any kind of application. Researchers would like to much effort that have made to control micro and mesopores in carbon materials, and prepare them in achieving the final goal. The presence of mesopores in electrodes based on CNTs, due to the central canal and entanglement enables easy access of ions from electrolyte. For electrodes built from multiwalled carbon nanotubes (MWCNTs), specific capacitance in a range of 4–135 F/g was found. For single-walled carbon nanotubes (SWCNTs), a maximum specific capacitance of 180 F/g is reported. A comparative investigation of the specific capacitance achieved with CNTs and activated carbon material reveals the fact activated carbon material exhibited significantly higher capacitance. Supercapacitor CNTs-based electrodes were fabricated by direct synthesis of nanotubes on the bulk Ni substrates, by means of plasma enhanced chemical vapor deposition of methane and hydrogen. The specific capacitance of electrodes with such nanotubes was of 49 F/g. MWCNTs were electrochemically oxidized and their performance in EDLCs was studied [45–64–68].

5.5   CARBON NANOTUBE: PROPERTIES AND APPLICATION

An article by Iijima that showed that carbon nanotubes are formed during arc-discharge synthesis of C_{60}, and other fullerenes also triggered an outburst of the interest in carbon nanofibers and nanotubes. These nanotubes may be even single-walled, whereas low-temperature, catalytically grown tubes are multiwalled. It has been realized that the fullerene-type materials and the carbon nanofibers known from catalysis are relatives, and this broadens the scope of knowledge and of applications. This chapter describes the issues around application and production of carbon nanostructures. Electrospinning is a simple and versatile method for generating ultrathin fibers from a rich variety of materials that include polymers, nanocomposites, and ceramics. In a typical process, an electrical potential is applied between a droplet of a polymer solution, or melt, held at the end of a capillary tube and a grounded target. When the applied electric field overcomes the surface tension of the droplet, a charged jet of polymer solution is ejected.

The following parameters and processing variables affect the electrospinning process: (i) system parameters such as molecular weight, molecular weight distribution and architecture (branched, linear, etc.) of the polymer, and polymer solution properties (viscosity, conductivity, dielectric constant, and surface tension, charge carried by the spinning jet) and (ii) process parameters such as electric potential, flow rate and concentration, distance between the capillary and collection screen, ambient parameters (temperature, humidity and air velocity in the chamber) and finally motion of the target screen. Morphological changes can occur upon decreasing the distance between the syringe needle and the substrate. Increasing the distance or decreasing the electrical field decreases the bead density, regardless of the concentration of the polymer in the solution. Elemental carbon in the sp^2 hybridization can form a variety of amazing structures.

The nanotubes consisted of up to several tens of graphitic shells (so-alled multi-walled carbon nanotubes (MWNTs)) with adjacent shell separation of ~0.34 nm, diameters of ~1 nm and high length/diameter ratio. Two years later, Iijima and Ichihashi synthesized single-walled carbon nanotubes (SWNTs). There are two main types of carbon nanotubes that can have high structural perfection. SWNT consist of a single graphite sheet seamlessly wrapped into a cylindrical tube. WNTs comprise an array of such nanotubes that are concentrically nested like rings of a tree trunk [54–57].

Recent discoveries of fullerene, a zero-dimensional form of carbon and carbon nanotube, which is a one-dimensional form, have stimulated great interest in carbon materials overall. Fullerenes are geometric cage-like structures of carbon atoms that are composed of hexagonal and pentagonal faces. When a bucky ball is elongated to form a long and narrow tube of few nanometers diameter approximately, which is the basic form of carbon nanotube. This stimulated a frenzy of activities in properties measurements of doped fullerenes. The discovery of fullerenes led to the discovery of carbon nanotubes by Iijima in 1991. The discovery of carbon nanotubes created much excitement and stimulated extensive research into the properties of nanometer scale cylindrical carbon networks.

Many researchers have reported mechanical properties of carbon nanotubes that exceed those of any previously existing materials. Although there are varying reports in the literature on the exact properties of carbon nanotubes, theoretical and experimental results have shown extremely high modulus, greater than 1 TPa (the elastic modulus of diamond is 1.2 TPa) and reported strengths 10–100 times higher than the strongest steel at a fraction of the weight. Indeed, if the reported mechanical properties are accurate, carbon nanotubes may result in an entire new class of advanced materials [58–62].

In addition to the exceptional mechanical properties associated with carbon nanotubes, they also possess superior thermal and electric properties such as thermally stable up to 2,800°C in vacuum, thermal conductivity about twice as high as diamond, electric current carrying capacity 1,000 times higher than copper wires. These exceptional properties of carbon nanotubes have been investigated for devices such as field emission displays, scanning probe microscopy tips, and microelectronic devices. Carbon nanotubes present significant opportunities to basic science and nanotechnology, and pose significant challenge for future work in this field. The approach of direct growth of nanowires into ordered structures on surfaces is a promising route to approach nanoscale problem and create novel molecular scale devices with advanced electrical, electromechanical and chemical functions [54].

5.5.1 CNT COMPOSITE-BASED ADSORBENTS

Potential practical applications of CNTs such as chemical sensors, field emission, electronic devices, high sensitivity nanobalance for nanoscopic particles, nanotweezers, reinforcements in high-performance composites, biomedical and chemical investigations, anode for lithium-ion batteries, supercapacitors, and hydrogen storage have been reported. Even though the challenges in fabrication may prohibit realization of many of these practical device applications, the fact that the properties of CNTs can be altered by suitable surface modifications can be exploited for more imminent realization of practical devices. In this respect, a combination of CNTs and other nanomaterials, such as nanocrystalline metal oxide/CNTs, polymer/CNTs and metal filled CNTs may have unique properties and research have therefore been focused on the processing of these CNT-based nanocomposites and their different applications [54–57].

Adsorption of single metal ions, dyes and organic pollutants on CNT-based adsorbent composites is one of the most important applications of these materials. MWCNT/iron oxide magnetic composites were prepared and used for adsorptions of several metal ions. The CNTs were purified by using nitric acid which results in modification of the surface of the nanotubes with oxygen containing groups like carbonyl and hydroxyl groups. The adsorption capability of the composite is higher than that of nanotubes and activated carbon. The sorption of ions such as Pb(II) and Cu(II) ions on the composite were spontaneous and endothermic processes based on the thermodynamic parameters (ΔH, ΔS, and ΔG) calculated from temperature dependent sorption isotherms. Alumina-coated MWCNTs were synthesized and reported for its utilization as adsorbent for the removal of lead ions from aqueous solutions in two modes. With an increase in influent pH between 3 and 7, the percentage of lead removed increases. The adsorption capacity increases by increasing agitation speed, contact time and adsorbent dosage. The reported composite can be regenerated as it was confirmed by SEM and EDX analysis [58–60].

CNTs can be combined with various metal oxides for the degradation of some organic pollutants too. Carbon nanotubes/metal oxide (CNT/MO) composites can be prepared by various methods such as wet chemical, sol gel, physical and mechanical methods. To form nanocomposite, CNTs can be combined with various metal oxides like Ti_2O_3, ZnO, WO_3, Fe_2O_3, and Al_2O_3. The produced nanocomposite can be used for the removal of various pollutants. Nanoscale Pd/Fe particles were combined with MWNTs and the resulted composite was used to remove 2,4-dichlorophenol (2,4-DCP). It was reported that the MB adsorption was pH-dependent and adsorption kinetics was best described by the pseudo-second-order model. Iron oxide/CNT composite was reported to be efficient adsorbent for remediation of chlorinated hydrocarbons. The efficiency of some other nanocomposites such as CNT/alumina, CNT/titania and CNT/ZnO has also been reported [60–62].

Most important CNT-based as adsorbent composites include: CNT–chitosan composites, CNT–ACF, CNTs–Fe_3O_4, CNTs–dolomite, CNTs–cellulose, and CNTs–graphene. Chitosan (CS) is one of the best adsorbents for the removal of dyes due to its multiple functional groups, biocompatibility and biodegradability, but its low mechanical strength limits its commercial applications. Impregnation of CS hydrogel beads with CNTs (CS/CNT beads) resulted in significant improved mechanical strength. In CS/CNT composite, CNTs and CS are like a symbiosis, CNTs help to improve the mechanical strength of CS, whereas CS helps reduce the cost of CNTs for adsorption, while the resulted composite solves the problem of separating CNTs from aqueous medium. To resolve the aggregation and dispersion problem of CNTs, prepared the CNTs/activated carbon fabric (CNTs/ACF) composite and its application was investigated for the removal of phenol and basic violet 10 (BV10). CNTs/ACF was prepared via directly growing nanoscaled CNTs on microscaled carbon matrix. Polyacrylonitrile was used as a source of carbon. From the results, it was observed that dye adsorption equilibrium time for CNTs/ACF is shorter as compared to ACF and monolayer adsorption capacity does not display a linear increase with increasing the BET surface area.

The decoration of CNTs tends to lower the porosity of the ACF from 1,065 to 565 m^2/g. The adsorption of BV10 onto ACF and CNTs/ACF was 162.4 and 220 mg/g, respectively. This finding indicates that the total microporosity of ACF cannot be fully accessed by the dye molecules. Therefore, the appearance of CNTs plays a positive role in (i) facilitating the pore accessibility to adsorbates and (ii) providing more adsorptive sites for the liquid-phase adsorption. This reflects that CNTs/ACF contains a large number of mesopore channels, thus preventing the pore blockage from the diffusion path of micropores for adsorbates to penetrate [72–73].

Incorporation of magnetic property in CNTs is another good technique to separate CNTs from solution. The magnetic adsorbent can be well dispersed in the water and easily separated magnetically. Magnetic-modified MWCNTs were used for removal of cationic dye crystal violet (CV), thionine (Th), janus green B (JG), and methylene blue (MB). To find the optimum adsorption, effect of various parameters including initial pH, dosages of adsorbent and contact time have been investigated. The optimum adsorption was found to be at pH 7.0 for all dyes. The removal efficiency of cationic dyes using GG/ MWCNT/Fe_3O_4 is higher as compared with other adsorbents such as MWCNTs and MWCNT/Fe_3O_4.

The magnetic GG/MWCNT/Fe_3O_4 possesses the high adsorption properties and magnetic separation and can therefore be used as magnetic adsorbents to remove the contaminants from aqueous solutions. A novel magnetic composite bioadsorbent composed of chitosan wrapping magnetic nanosized γ-Fe_2O_3 and MWCNTs (m-CS/γ-Fe_2O_3/MWCNTs) was prepared for the removal of methyl orange. The adsorption capacity of MO onto m-CS/γ-Fe_2O_3/ MWCNTs was 2.2 times higher than m-CS/γ-Fe_2O_3. The adsorption capacity of MO onto m-CS/γ-Fe_2O_3/MWCNTs was also higher than MWCNTs. Kinetics data and adsorption isotherm data were better fitted by pseudo-second-order kinetic model and by Langmuir isotherm, respectively. The researchers think CS is responsible for the exothermic adsorption process because with the increase in the temperature, polymeric network of CS changed/deshaped, which reduced the porosity of the biosorbent and hindered the diffusion of dye molecules at high temperature [58–62–72–73].

Recently, a novel adsorbent was developed by inserting MWCNTs into the cavities of dolomite for scavenging of ethidium the foam line CNTs/dolomite adsorbent. Foam-like ternary composite PUF/diatomite/dispersed-MWCNTs, gave the highest capacities for adsorption of these dyes, followed by PUF/agglomerated-MWCNTs, and then PUF/dispersed-MWCNTs. Adsorption isotherm study revealed the monolayer adsorption at higher concentration and multilayer adsorption at lower concentration. Pseudo first-order kinetics gives the best fitted results compared to the pseudo second order.

From the results, it is evident that carboxylic group on the adsorbent plays the important role for the removal of MB as ionized to COO− at higher pH and bind with MB through electrostatic force. Self-assembled cylindrical graphene–MCNT (G–MCNT) hybrid, synthesized by the one pot hydrothermal process was used as adsorbent for the removal of methylene blue in batch process. G–MCNT hybrid showed good performance for the removal of MB from aqueous solution. The adsorption capacity of G–MCNTs is much higher than MCNTs. Therefore, G–CNTs hybrid could be utilized as an efficient adsorbent for environmental remediation [72–73].

5.6   CARBON NANFIBER: PROPERTIES AND APPLICATION

Carbon nanofibers (diameter range, 3–100 nm) have been known for a long time as a nuisance that often emerges during catalytic conversion of carbon-containing gases. The recent outburst of interest in these graphitic materials originates from their potential for unique applications as well as their chemical similarity to fullerenes and carbon nanotubes. This chapter focused on the growth of nanofibers using metallic particles as a catalyst to precipitate the graphitic carbon. First, it summarized some of the earlier literature that has contributed greatly to understand the nucleation and growth of carbon nanofibers and nanotubes. Thereafter, it described in detail recent progress to control the fiber surface structure, texture, and growth into mechanically strong agglomerates. It is argued that carbon nanofibers are unique high surface area materials (200 m1/g) that can expose exclusively either basal graphite planes or edge planes. It is shown that the graphite surface structure and the lyophilicity play a crucial role during metal emplacement and catalytic use in liquid phase catalysis.

An article by Iijima showed that carbon nanotubes are formed during arc-discharge synthesis of C_{60}, and other fullerenes also triggered an outburst of the interest in carbon nanofibers and nanotubes. These nanotubes may be even single walled; whereas, low-temperature, catalytically grown tubes are multiwalled. It has been realized that the fullerene type materials and the carbon nanofibers known from catalysis are relatives and this broadens the scope of knowledge and of applications. It has been realized; however, that arc-discharge and laser-ablation methods lead to mixtures of carbon materials and thus to a cumbersome purification to obtain nanofibers or nanotubes [88].

From an application viewpoint, some of best application of carbon nanofibers include carbon nanofibers as catalyst support materials, carbon nanofiber-based electrochemical biosensors, CNF-based oxidase biosensors, CNF-based immunosensor and cell sensor and hydrogen storage. The overall economics are affected by the fiber yield, the feedstock used, the rate of growth, and the reactor technology [88–92].The growth of parallel fibers using iron as the catalyst has been studied in detail by high-resolution transmission electron microscopy (HR-TEM). It is noted that the graphite layers grow at an angle iron surface, thus leading to parallel fibers. The diameter of the fibers can be varied by variation of the metal particle size. If we want to vary the fiber diameter for a macroscopic sample, however, we need a narrow metal particle-size distribution. In general, one can say that the fibers do not contain micropores and that the surface area can range from 10 to 200 m^2/g and the mesopore volume ranges between 0.50 and 2.0 mL/g. Note that these pore-volume data are obtained within fibers as grown, specific treatments in the liquid phase can be applied to largely reduce the pore volume and to obtain much denser and compact fiber structures.

Compared to the large volume of literature on the mechanism of growth, the studies on the macroscopic, mechanical properties of bodies consisting of agglomerates of carbon nanofibers have been limited in number give a useful description of the tertiary structures that can be obtained; that is, "bird nests," nepo "net," and "combedyarn." In general, porous bodies of carbon nanofibers are grown from porous supported metal catalyst bodies. Some others in the size range of micrometer to millimeter. As carbon precursors, PAN and pitches were frequently used, probably because both of them are also used in the production of commercial carbon fibers.

In addition, poly (vinyl alcohol) (PVA), polyimides (PIs), polybenzimidazol (PBI) poly (vinylidene fluoride) (PVDF), phenolic resin and lignin were used. In order to convert electrospun polymer nanofibers to carbon nanofibers, carbonization process at around 1,000°C has to be applied. In principle, any polymer with a carbon backbone can potentially be used as a precursor. For the carbon precursors, such as PAN and pitches, so-called stabilization process before carbonization is essential to keep fibrous morphology, of which the fundamental reaction is oxidation to change resultant carbons difficult to be graphitized at high temperatures as 2,500°C [89–90].

Carbon is an important support material in heterogeneous catalysis, in particular for liquid-phase catalysis. A metal support interaction between Ru and C was suggested as a possible explanation for these very interesting observations. More recent work focuses on the use of platelet type fibers exposing exclusively graphite edge sites. Using a phosphorus-based treatment, preferential blocking of so-called armchair faces occurs. Deposition of nickel onto the thus modified CNF enabled one to conclude that the nickel particles active for hydrogenation of light alkenes reside on the zigzag faces. More characterization work is needed to substantiate these interesting claims. Others have carried out, by far, the most extensive work on CNF as carbon support material. A driving force for exploring CNF supports was related to the replacement of active carbon as support for liquid phase catalysis. For the CNF support, no shift of the PSD is apparent, whereas with an activated carbon (AC) support, severe attrition is apparent.

They compare the PSD after ultrasonic treatment of the CNF and of the AC support. Clearly, AC displays a much broader PSD with, moreover, a significant number of fines. Nanocomposite electrodes made of carbon nanofibers and paraffin wax were characterized and investigated as novel substrates for metal deposition and stripping processes. Since CNFs have a much larger functionalized surface area compared to that of CNTs, the surface active groups to volume ratio of these materials is much larger than that of the glassy-like surface of CNTs. This property, combined with the fact that the number and type of functional groups on the outer surface of CNFs can be well controlled, is expected to allow for the selective immobilization and stabilization of biomolecules such as proteins, enzymes, and DNA. Also, the high conductivity of CNFs seems to be ideal for the electrochemical transduction. Therefore, these nanomaterials can be used as scaffolds for the construction of electrochemical biosensors [64–68].

Compared with conventional ELISA-based immunoassays, immunosensors are of great interest because of their potential utility as specific, simple, label-free and direct detection techniques and the reduction in size, cost, and time of analysis. Due to its large functionalized surface area and high surface active groups-to-volume ratio. Hydrogen storage is an essential prerequisite for the widespread deployment of fuel cells, particularly in transport. Huge hydrogen storage capacities, up to 67 percent, were reported. Unfortunately such astonishing values could not confirmed by other research teams worldwide.. Some reviews provides basics of hydrogen storage on carbon materials, the types of carbon materials with potential for hydrogen storage, the measured hydrogen storage capacities of these materials, and based on calculations, an approximation of the theoretical achievable hydrogen storage capacity of carbon materials [91–92].

5.7 ACTIVATED CARBON NANOTUBE: PROPERTIES AND APPLICATION

Electrospinning, a simple approach to make very fine fibers ranging from nano-to-micro scales, is attracting more attention due to the high porosity and high surface area to volume ratio of electrospun membranes. These properties contribute to potential applications of electrospun membranes in carbon and graphitic nanofiber manufacturing, tissue scaffolding, drug delivery systems, filtration and reinforced nanocomposites. The researchers also tried poly (amic acid) as a precursor to make activated carbon nanofibers. These studies employed physical activation to produce pores in precursor fibers. Compared with physical activation, the chemical activation process has important advantages, including low heat treatment temperature, short period of processing time, large surface area and high carbon yield; however, there has been no work reported for chemically activated carbon nanofibers from electrospun PAN. The effects of electrospinning variables, such as applied voltage, pump flow rate, and distance between the needle tip and collector, on the resulting nanofiber diameters were studied. Mechanical properties, such as tensile, tear and burst strength, of electrospun PAN nonwoven membranes were measured and the quantitative relationships between membrane thickness and these properties were established [93–94–95].

Other physical properties, such as air permeability, interfiber pore size and porosity, were also studied. Activated carbon nanofibers were produced from electrospun PAN by chemical activation with potassium hydroxide (KOH) as the activating agent. They were characterized by morphology, Fourier transform infrared spectroscopy (FTIR), Brunauer–Emmett–Teller

(BET) surface area, total pore volume and pore size distribution. There are two processes for manufacturing the carbon nanofiber (CNF), namely, the vapor-grown approach and the polymer spinning approach. The activated carbon nanofiber (ACNF) is the physically or chemically activated CNF, which have been, in many investigations, practically applied in electric double layer capacitors, organic vapor recovery, catalyst support, hydrogen storage, and so on. In practice, the physical activation method involves carbonizing the carbon precursors at high temperatures and then activating CNF in an oxidizing atmosphere such as carbon dioxide or steam [96–97].

The chemical activation method involves chemically activating agents such as alkali, alkaline earth metals, some bases such as potassium hydroxide (KOH) and sodium hydroxide, zinc chloride, and phosphoric acid (H_3PO_4). In essence, most chemical activation on CNF used KOH to get highly porous structure and higher specific surface area. Unfortunately, large amount of solvent were needed to prepare the polymer solution for electrospinning and polymer blend, causing serious environmental problem thereafter. A series of porous amorphous ACNF were studied. Utilizing the core/shell microspheres that were made of various polymer blends with solvent. In their approach, the phenol formaldehyde-derived CNF were chemically activated by the alkaline hydroxides, and the thus-prepared ACNF were applied as super-capacitor electrodes and hydrogen storage materials [98–99–100].

In a continuous effort, some researchers proceed to investigate and compare the various chemical activation treatments on the CNF thus prepared, with particular emphasis on the qualitative description and quantitative estimation on the surface topology by AFM, and their relation to the microstructure of ACNF. PAN fiber following the spinning process by several ways such as modification through coating, impregnation with chemicals (catalytic modification) and drawing/stretching with plasticizer. The post spinning modifications indirectly affect and ease the stabilization in several ways such as reducing the activation energy of cyclization, decreasing the stabilization exotherm, increasing the speed of cyclization reaction, and also improving the orientation of molecular chains in the fibers.

One of the well-known posts spinning treatment for PAN fiber precursor is modification through coatings. The PAN fibers are coated with oxidation resistant resins such as lubricant (finishing oil), antistatic agents, and emulsifiers which are basically used as spin finish on the precursor fiber. Coating with certain resins also acts in the same manner as the co-monomer in reducing the cyclization exotherm thus improving the mechanical properties of the resulting carbon fibers. Due to their excellent lubricating properties, silicone based compounds are mostly used as the coating material for PAN precursor fibers. Tensile load and tear strength of electrospun PAN membranes increased with thickness, accompanied with a decrease in air permeability; however, burst strength was not significantly influenced by the thickness [101–102–103].

Electrospun PAN nanofiber membranes were stabilized in air and then activated at 800°C with KOH as the activating agent to make activated carbon nanofibers. Stabilized PAN membranes showed different breaking behaviors from those before stabilization. The activation process generated micropores which contributed to a large surface area of 936.2 m^2/g and a micropore volume of 0.59 cc/g. Pore size distributions of electrospun PAN and activated carbon nanofibers were analyzed based on the Dubinin–Astakhov equation and the generalized Halsey equation. The results showed that activated carbon nanofibers had many more micropores than electrospun PAN, increasing their potential applications in adsorption. Based on a novel solvent-free co-extrusion and melt-spinning of polypropylene/(phenol formaldehyde polyethylene) based core/sheath polymer blends, a series of activated carbon nanofibers (ACNFs) have been prepared and their morphological and microstructure characteristics analyzed by scanning electron microscopy, atomic force microscopy (AFM), Raman spectroscopy, and X-ray diffractometry with particular emphasis on the qualitative and quantitative AFM analysis. Post spinning treatment of the current commercial PAN fiber based on the author's knowledge, reports on the post spinning modification process of the current commercial fiber are still lacking in the carbon fiber manufacturers' product data sheet, which allows us to assume that the current commercial carbon fibers still do not take full advantage of any of these treatments yet. During stabilization and carbonization of polymer nanofibers, they showed significant weight loss and shrinkage, resulting in the decrease of fiber diameter [104–107].

From an application viewpoint, Some of best application of carbon nanofibers include: ACNF as anodes in lithium-ion battery, Organic removal from waste water using, ACNF as cathode catalyst or as anodes for microbial fuel cells (MFCs), Electrochemical properties of ACNF as an electrode for supercapacitors, Adsorption of some toxic industrial solutions and air pollutants on ACNF [108–120].

Activated carbon nanofibers (ACNFs) with large surface areas and small pores were prepared by electrospinning and subsequent thermal and chemical treatments. These activated CNFs were examined as anodes for lithium-ion batteries(UBs) without adding any non-active material. Their electrochemical behaviors show improved lithium-ion storage capability and better cyclic stability compared with unactivated counterparts. The development of high-performance rechargeable lithium-ion batteries (LIBs) for efficient energy storage has become one of the components in today's information rich mobile society [114].

Microbial fuel cell (MFC) technologies are an emerging approach to wastewater treatment. MFCs are capable of recovering the potential energy present in wastewater and converting it directly into electricity. Using MFCs may help offset wastewater treatment plant operating costs and make advanced wastewater treatment more affordable for both developing and industrialized nations. In spite of the promise of MFCs, their use is limited by low power generation efficiency and high cost. Some researchers conclude that the biggest challenge for MFC power output lies in reactor design combining high surface area anodes with low ohmic resistances and low cathode potential losses. Power density limitations are typically addressed by the use of better suited anodes, use of mediators, modification to solution chemistry or changes to the overall system design. Employing a suitable anode, however, is critical since it is the site of electron generation.

An appropriately designed anode is characterized by good conductivity, high specific surface area, biocompatibility and chemical stability. Anodes currently in use are often made of carbon and/or graphite. Some of these anodes include but are not limited to: graphite plates/rods/felt, carbon fiber/cloth/foam/paper and reticulated vitreous carbon (RVC). Carbon paper, cloth and foams are among the most commonly used anodes and their use in MFCs has been widely reported employ activated carbon nanofibers (ACNF) as the novel anode material in MFC systems. Compared with other activated carbon materials, the unique features of ACNF are its mesoporous structure, excellent porous interconnectivity, and high bioavailable surface area for biofilm growth and electron transfer [115].

Among the diverse carbonaceous adsorbents, activated carbon fiber (ACF) is considered to be the most promising due to their abundant micropores, large surface area, and excellent adsorption capacity. Therefore, the investigation of formaldehyde adsorption has been steadily conducted using ACFs. However, most of precedent works have generally concerned about removal of concentrated formaldehyde in aqueous solution (formalin, and thus there was limited information whether these materials could be used in the practical application, because the concentration of formaldehyde in indoor environment was generally very low (below 1 ppm). The nitrogen containing functional groups in ACF played an important role in increasing formaldehyde adsorption ability, as also described elsewhere. However, the PAN-based ACFs still have problems in a practical application, because the adsorption capability is drastically reduced under humid condition [116–120].

5.8   PORE FORMATION IN CARBON MATERIALS, PORE CHARACTERIZATION AND ANALYSIS OF PORES

About pore formation in carbon materials, this is accepted that all carbon materials, except highly oriented graphite, contain pores, because they are polycrystalline and result from thermal decomposition of organic precursors. During their pyrolysis and carbonization, a large amount of decomposition gases is formed over a wide range of temperatures, the profile of which depends strongly on the precursors. Since the gas evolution behavior from organic precursors is strongly dependent on the heating conditions, such as heating rate, pressure, etc., the pores in carbon materials are scattered over a wide range of sizes and shapes. These pores may be classified as shown in Table 5.3 [1–2–3].

TABLE 5.3   Classification of pore formation in carbon materials

(1) Based on their origin	(2) Based on their size	(3) Based on their state		
Intraparticle pores	Intrinsic intraparticle pores			
	Extrinsic intraparticle pores	Micropores	< 2 nm	Open pores

TABLE 5.3   *(Continued)*

Interpar-ticle pores	Rigid interpar-ticle pores Flexible interpar-ticle pores	Mesopores	2~50 nm	Ultramicro-pores <0.7 nm	Closed pores (latent pores)
		Macropores	> 50 nm	Supermi-cropores 0.7–2 nm	

Table 5.3 shows that based on their origin, the pores can be categorized into two classes, intraparticle and interparticle pores. The intraparticle pores are further classified into two, intrinsic and extrinsic intraparticle pores. The former class owes its origin to the crystal structure, that in most activated carbons, large amounts of pores of various sizes in the nanometer range are formed because of the random orientation of crystallites; these are rigid interparticle pores.

A classification of pores based on pore sizes was proposed by the International Union for Pure and Applied Chemistry (IUPAC). As illustrated in Table 5.3, pores are usually classified into three classes: macropores (>50 nm), mesopores (2–50 nm) and micropores (<2 nm). Micropores can be further divided into supermicropores (with a size of 0.7–2 nm) and ultramicropores (<0.7 nm in size). Since nanotechnology attracted the attention of many scientists recently, the pore structure has been required to be controlled closely. When scientists wanted to express that they are controlling pores in the nanometer scale, some of them preferred to call the smallest pores nanosized pores, instead of micro/mesopores [1–2].

Pores can also be classified on the basis of their state, either open or closed. In order to identify the pores by gas adsorption (a method which has frequently been used for activated carbons), they must be exposed to the adsorbate gas. If some pores are too small to accept gas molecules they cannot be recognized as pores by the adsorbate gas molecules. These pores are called latent pores and include closed pores. Closed pores are not necessarily in small size. Pores in carbon materials have been identified by different techniques depending mostly on their sizes. Pores with nanometer sizes, that is, micropores and mesopores, are identified by the analysis of gas adsorption isotherms, mostly of nitrogen gas at 77 K [41–46].

The basical theories, equipments, measurement practices, analysis procedures and many results obtained by gas adsorption have been reviewed in different publications. For macropores, mercury porosimetry has been frequently applied. Identification of intrinsic pores, the interlayer space between hexagonal carbon layers in the case of carbon materials, can be carried out by X-ray diffraction (XRD). Recently, direct observation of extrinsic pores on the surface of carbon materials has been reported using microscopy techniques coupled with image processing techniques, namely scanning tunneling microscopy (STM) and atomic force microscopy (AFM) and transmission electron microscopy (TEM) for micropores and mesopores, and scanning electron microscopy (SEM) and optical microscopy for macropores [1–3].

The most important pore characterization methods include scanning tunneling microscopy (STM), atomic force microscopy (AFM), transmission electron microscopy (TEM), gas adsorption, calorimetric methods, small-angle X-ray scattering (SAXS), small-angle neutron scattering (SANS), positron annihilation lifetime spectroscopy (PALS), scanning electron microscopy (SEM), optical microscopy, mercury porosimetry, and molecular resolution porosimetry. Adsorption from solution using macromolecules has been applied to macropore analysis, but we still need more examinations. It is difficult to compare one adsorption isotherm with another, but determination of the deviation from the linearity using a standard adsorption isotherm is accurate. The plot constructed with the aid of standard data is called a comparison plot. The representative comparison plots are the t and alpha plots [41–42].

The molecular adsorption isotherm on nonporous solids can be well described by the BET theory. The deviation from the linearity of the t plot gives information on the sort of pores, the average pore size, the surface area, and the pore volume. However, the t plot analysis has the limited applicability to the microporous system due to the absence of explicit monolayer adsorption. The construction of the alpha plot does not need the monolayer capacity, so that it is applicable to microporous solids.

The straight line passing the origin guarantees multilayer adsorption, that is, absence of meso and/or micropores; the deviation leads to valuable information on the pore structures. A nonporous solid has a single line passing the origin, while the line from the origin for the alpha plot of the mesoporous system bends. The slopes of the straight line through the origin and the line at the high as region gives the total and mesoporous surface areas, respectively. The type of the alpha plot suggests the presence of ultramicropores and/or supermicropores. Detailed analysis results will be shown for the micropore analysis [43–46].

In the discussion of the mesopore shape, the contact angle, is assumed to be zero (uniform adsorbed film formation). The lower hysteresis loop of the same adsorbate encloses at a common relative pressure depending to the stability of the adsorbed layer regardless of the different adsorbents due to the so called tensile strength effect. This tensile strength effect is not sufficiently considered for analysis of mesopore structures. The Kelvin equation provides the relationship between the pore radius and the amount of adsorption at a relative pressure. Many researchers developed a method for the calculation of the pore size distribution on the basis of the Kelvin equation with a correction term for the thickness of the multilayer adsorbed film.

They so called BJH (Barret–Joyer–Halenda) and DH (Dollimore–Heal) methods have been widely used for such calculations. However, in other articles, only a simple Fortran program for the DH method is shown. (This program can be easily used for the analysis of the mesopore size distribution). The thickness correction is done by the Dollimore–Heal equation. One can calculate the mesopore size distribution for cylindrical or slit-shaped mesopores with this program. Therefore, the adsorption branch provides more reliable results. However, the adsorption branch gives a wide distribution compared to the desorption branch due to gradual uptake. Theoretical studies on these points are still done [133].

The pore size distribution from the Kelvin equation should be limited to mesopores due to the ambiguity of the meniscus in the microporous region. It is well known that the presence of micropores is essential for the adsorption of small gas molecules on activated carbons. However, when the adsorbate is polymer, dye or vitamin, only mesopores allow the adsorption of such giant molecules and can keep even bacteria. The importance of mesopores has been pointed out not only for the giant molecule adsorption, but also for the performance of new applications such as electric double layer capacitors. Thus, the design and control of mesoporosity is very desirable both for the improvement of performance of activated carbon and for the development of its new application fields [1–3].

Important parameters that greatly affect the adsorption performance of a porous carbonaceous adsorbent are porosity and pore structure. Consequently, the determination of pore size distribution (PSD) of carbon nanostructures adsorbents is of particular interest. For this purpose, various methods have been proposed to study the structure of porous adsorbents. A direct but cumbersome experimental technique for the determination of PSD is to measure the saturated amount of adsorbed probe molecules which have different dimensions.

However, there is uncertainty about this method because of networking effects of some adsorbents including activated carbons and carbon nanostructures. Other experimental techniques that usually implement for characterizing the pore structure of porous materials are mercury porosimetry, X-ray diffraction (XRD) or small angle X-ray scattering (SAXS), and immersion calorimetry.

A large number of simple and sophisticated models have been presented to obtain a realistic estimation of PSD of porous adsorbents. Relatively simple but restricted applicable methods such as Barret, Joyner, and Halenda (BJH), Dollimore and Heal (DH), Mikhail et al., (MP), Horvath and Kawazoe (HK), Jaroniec and Choma (JC), Wojsz and Rozwadowski (WR), Kruk–Jaroniec–Sayari (KJS), and Nguyen and Do (ND) were presented from 1951 to 1999 by various researchers for the prediction of PSD from the adsorption isotherms [133–139].

For example, the BJH method which is usually recommended for mesoporous materials is in error even in large pores with dimension of 20 nm. The main criticism of the MP method, in addition to the uncertainty regarding the multilayer adsorption mechanism in micropores, is that we should have a judicious choice of the reference isotherm. HK model was developed for calculating micropore-size distribution of slit-shaped pore; however, the HK method suffers from the idealization of the micropore filling process. Extension of this theory for cylindrical and spherical pores was made by Saito and Foley and Cheng and Ralph. By applying some modifications on the HK theory, some improved models for calculating PSD of porous adsorbents have been presented. Gauden et al. extended the Nguyen and Do method for the determination of the bimodal PSD of various carbonaceous materials from a variety of synthetic and experimental data. The pore range of applicability of this model besides other limitations of ND method is its main constraint.

In 1985, Bunke and Gelbin determined the PSD of activated carbons based on liquid chromatography (LC) [133]. Choice of suitable solvent and pore range of applicability of this method are two main problems that restrict its general applicability. More sophisticated methods such as molecular dynamics (MD), Monte Carlo simulation, grand canonical Monte Carlo simulations (GCMS), and density functional theory (DFT) are theoretically capable of describing adsorption in the pore system. The advantages of these methods are that they can apply on wide range of pores width. But, they are relatively complicated and provide accurate PSD estimation based on just some adsorbates with specified shapes [140–145].

5.9 RECENTLY STUDY WORKS ABOUT CONTROLLING PORE SIZE

Summary of some recently reported papers about confirmal modeling of controlling of pore size in carbon-based nanoadsorbent are presented in Table 5.4 [8–40–121–145].

TABLE 5.4 Summary of recently confirmal models for controlling of pore size in carbon based nano adsorbents

Carbon material type	Applied model and simulation methods	References
AC, ACF, MSC	–	T.k et al., (2000) [8]
Carbon structures	–	K.k et al., (1994) [9]
Ac	G CMC and DFT (Monte Carlo simulations)	j.p.o et al., (1998) [10]
Carbon structures	N_2 and SIMULATION NONE LOCAL THEORY model	Ch.L et al., (1993) [11]
Ac	G CMC	D.C et al., (2005) [12]
ACF	H_2S adsorption	W.F et al., (2005) [13]
C-ZIFs	–	T.R.B et al., (2009) [14]
Carbon structures	ND and DFT	P.k et al., (2003) [15]

TABLE 5.4   *(Continued)*

Carbon material type	Applied model and simulation methods	References
Ac	N$_2$ and T-plot	Ru. L.T et al., (2005) [16]
Ac	Montecarlo simulations	P.K et al., (2008) [17]
p-carbon	NLDFT-BjH method	A.p et al., (2007) [18]
ACF and AAPFs	DR equation	C.L.M et al., (1998) [19]
AAC	–	M.E. et al., (2001) [20]
ACH structure	–	K.g et al., (2000) [21]
ACF	–	C.P et al., (2000) [22]
Ac	–	W.M.D et al., (2000) [23]
ACF	–	C.P et al., (2001) [24]
AC	IAST–freundlich model	K.E et al., (2001) [25]
ACF	–	Y.L.M et al., (2001) [26]
MSC, AC	–	R.F.P.M et al., (2001) [27]
P-ACS	–	J.B-Y et al., (2002) [28]
AC	GCMS and SIE equation—methane adsorption	D.C et al., (2002) [29]
AC	–	X.P et al., (2005) [30]
Carbon structures	–	O.T et al., (2003) [31]
Carbon structures	–	J.Z et al., (2007) [32]

TABLE 5.4 *(Continued)*

Carbon material type	Applied model and simulation methods	References
Carbide-derived carbons	–	J.Ch et al., (2006) [33]
C-xerogels	–	C.L et al., (99) [34]
Carbide-derived carbons CDCs	–	Y.G et al., (2003) [35]
Carbon black	NLDFT	E.A.U et al., (2006) [36]
Carbon Structures	–	S.H et al., (2003) [37]
Glassy Carbon	Monte carlo simulation nadsorption	M P.M et al., (2006) [38]
Carbon Structures	DFT–Ar adsorption at 77 K	R.J.D et al., (2000) [39]
AC	N_2 adsorption-and DR equation–BjH	N. R K et al., (2000) [40]
P-Asc	N_2 adsorption at 77 K –BjH method	J.B.Y et al., (2002) [121]
Acf	DRS equation-N_2 adsorption at 150 C	C.L.M et al., (2000) [122]
AC	DFT–N_2 adsorption at 77 K	G.G-Z et al., (2009) [123]
AC	DS-HK-IHK	A.O et al., (2012) [124]
CNT composite	Molecular dynamics simulation (MDS)–PMF	Zhijun et al., [125] (2009)
CNT/SIO$_2$	MDS-AIREBO	Ong et al., [126] (2010)
CNT	MDS–PEOE algorithm-PME	Liu et al., [127] (2009)

TABLE 5.4 *(Continued)*

Carbon material type	Applied model and simulation methods	References
CNT/PE composite	MDS–Brenner–Newtonian equation	Zhang et al., [128] (2011)
SW/CNT	MDS–quantum potential–GCMC	Irle et al., [129] (2009)
SW/CNT	MDS–CPMD–DFT	Zang et al., [130] (2009)
CNT/Polymer	MDS–SHAKE algorithm–DL–Poly	Frankland et al., [131] (2002)
SW/CNT	MDS–GCMC–LJ–LORENTS BERTHELOT	Frankland et al., [132] (2002)
CNT/polymer composite	MDS–NPT	Han et al., [133] (2007)
2D graphene	ADS–DFT–APMD–PAW	Lehtinen et al., [134] (2010)
SW/CNT	MDS–Brenner	Shigeo.M., [135] (2003)
CNT ropes	GCMC	Williams et al., [136] (2000)
CNT/sodium	MDS	Diao et al., [137] (2008)
CNT	Abintio, K. Sh-Hamil	Noel.[138] (2010)
CNT	MDS–USHER algorithm–LJ potential	D.Nicholls et al., [139] (2012)
Nanotube	MDS–PES–Verlet algorithm	A.Ribos et al., [140] (2009)
CNF	MDS–GRASP–RFF	F.Sanz.N et al., [141] (2010)
SW/CNT	MDS–DFT–B3LYP	Shibuta et al., [142] (2003)
CNT	MDS–LJ potential	Mao et al., [143] (1999)
Metal membranes	EMD–GCMC–LJ potential	Mao et al., [144] (2010)
CNT	MDS-LJ–TIP5P	A.T et al. [145] (2009)

The recently research works about controlling of pore structure in carbon materials that categorized as fallowing include: First, using the grand canonical Monte Carlo (GCMC) method, or other simulation methods to determine adsorption isotherms in Ar (87 K) or N_2 (77k) that are simulated for all the carbon sample structures to reach optimum condition in experimental works. Second, experimental works to obtained PSD curves that show samples structures maybe micro or mesoporous (with different ratio of micro/mesopores). Finally PSD curves are calculated using the Horvath–Kawazoe (HK), density functional theory (DFT), D-R method, Barrett–Joyner–Halenda (BJH) approaches and other mathematical methods and this model results with those predicted by the experimental work results compared and adapted, to prove selected mathematical model is significant and simulation that applied is verified [121–145].

KEYWORDS

- **Activated carbon fibers**
- **Activated carbon tailoring**
- **Carbon nanostructure-based adsorbent**
- **Controlling of pore size**
- **Modeling methods**
- **Pore-size distribution**

REFERENCES

1. Bandosz, T. J.; Activated Carbon Surfaces in Environmental Remediation. Academic Press; **2006,** 7.
2. Marsh, H.; and Rodriguez-Reinoso, F.; Activated Carbon. Elsevier Science Limited; **2006**.
3. Bach, M. T.; Impact of Surface Chemistry on Adsorption: Tailoring of Activated Carbon. University of Florida; **2007**.
4. Machnikowski, J.; et al., Tailoring Porosity Development in Monolithic Adsorbents Made of KOH-Activated Pitch Coke and Furfuryl Alcohol Binder for Methane Storage†. *Energy Fuels.* **2010**, *24(6),* 3410–3414.
5. Yin, C. Y.; M.K. Aroua, and W.M.A.W. Daud, Review of modifications of activated carbon for enhancing contaminant uptakes from aqueous solutions: *Separation and Purification Technol.* **2007,** *52(3),* 403–415.

6. Muniz, J.; J. Herrero, and A. Fuertes, Treatments to enhance the SO_2 capture by activated carbon fibres. *Appl. Catalysis B: Environ.* **1998**, **18***(1)*, 171–179.

7. Houshmand, A.; Daud, W. M. A. W.; and Shafeeyan, M. S.; Tailoring the surface chemistry of activated carbon by nitric acid: study using response surface method. *Bull. Chem. Soc. Japan.* **2011**. **84***(11)*, 1251–1260.

8. Kyotani, T.; Control of Pore Structure in Carbon: Carbon. **2000**, **38***(2)*, 269–286.

9. Kaneko, K.; Determination of pore size and pore size distribution: 1. Adsorbents and catalysts. *J. Membrane Sci.* **1994**, **96***(1)*, 59–89.

10. Olivier, J.P.; Improving the models used for calculating the size distribution of micropore volume of activated carbons from adsorption data. *Carbon.* **1998**. **36***(10)*, 1469–1472.

11. Lastoskie, C.; K.E. Gubbins, and N. Quirke, Pore size heterogeneity and the carbon slit pore: a density functional theory model. *Langmuir.* **1993**, **9***(10)*, 2693–2702.

12. Cao, D.; and J. Wu, Modeling the selectivity of activated carbons for efficient separation of hydrogen and carbon dioxide. *Carbon.* **2005**, **43***(7)*, 1364–1370.

13. Feng, W.; et al., Adsorption of hydrogen sulfide onto activated carbon fibers: effect of pore structure and surface chemistry. *Environ. Sci. Technol.* **2005**, **39***(24)*, 9744–9749.

14. Banerjee, R.; et al., Control of pore size and functionality in isoreticular zeolitic imidazolate frameworks and their carbon dioxide selective capture properties. *J. Am. Chem. Soc.* **2009**, **131***(11)*, 3875–3877.

15. Kowalczyk, P.; et al., Estimation of the pore size distribution function from the nitrogen adsorption isotherm. Comparison of density functional theory and the method of Do and co-workers. *Carbon.* **2003**, **41***(6)*, 1113–1125.

16. Tseng, R. -L.; and S.-K. Tseng, Pore structure and adsorption performance of the KOH-activated carbons prepared from corncob. *J. Colloid Interface Sci.* **2005**, **287***(2)*, 428–437.

17. Kowalczyk, P.; A. Ciach, and A.V. Neimark, Adsorption-induced deformation of microporous carbons: Pore size distribution effect. *Langmuir.* **2008**, **24***(13)*, 6603–6608.

18. Terzyk, A. P.; et al., How realistic is the pore size distribution calculated from adsorption isotherms if activated carbon is composed of fullerene-like fragments? *Phys. Chem. Chem. Phys.* **2007**, **9***(44)*, 5919–5927.

19. Mangun, C.; et al., Effect of pore size on adsorption of hydrocarbons in phenolic-based activated carbon fibers. Carbon. **1998**, **36***(1)*, 123–129.

20. Endo, M.; et al.; High power electric double layer capacitor (EDLC's); from operating principle to pore size control in advanced activated carbons. *Carbon Sci.* **2001**, **1***(3&4)*, 117–128.

21. Gadkaree, K.; and M.; Jaroniec, Pore structure development in activated carbon honeycombs. *Carbon.* **2000**, **38***(7)*, 983–993.

22. Pelekani, C.; and V.L. Snoeyink, Competitive adsorption between atrazine and methylene blue on activated carbon: the importance of pore size distribution. *Carbon.* **2000**, **38***(10)*, 1423–1436.

23. Daud, W. M. A. W.; W.S.W. Ali, and M.Z. Sulaiman, The effects of carbonization temperature on pore development in palm-shell-based activated carbon. *Carbon.* **2000**, **38***(14)*, 1925–1932.

24. Pelekani, C.; and V.L. Snoeyink, A kinetic and equilibrium study of competitive adsorption between atrazine and Congo red dye on activated carbon: the importance of pore size distribution. *Carbon.* **2001**, **39***(1)*, 25–37.

25. Ebie, K.; et al., Pore distribution effect of activated carbon in adsorbing organic micropollutants from natural water. *Water Res.* **2001**, **35***(1)*, 167–179.

26. Mangun, C.L.; et al., Surface chemistry, pore sizes and adsorption properties of activated carbon fibers and precursors treated with ammonia. *Carbon.* **2001**, **39***(12)*, 1809–1820.

27. Moreira, R.; H. Jose, and A. Rodrigues, Modification of pore size in activated carbon by polymer deposition and its effects on molecular sieve selectivity. *Carbon.* **2001**, **39***(15)*, 2269–2276.

28. Yang, J.-B., et al., Preparation and properties of phenolic resin-based activated carbon spheres with controlled pore size distribution. Carbon. **2002**, **40***(6)*, 911–916.

29. Cao, D.; et al., Determination of pore size distribution and adsorption of methane and CCl_4 on activated carbon by molecular simulation. *Carbon.* **2002**, **40***(13)*, 2359–2365.

30. Py, X.; A. Guillot, and B. Cagnon, Activated carbon porosity tailoring by cyclic sorption/decomposition of molecular oxygen. Carbon. **2003**, **41***(8)*, 1533–1543.

31. Tanaike, O.; et al., Preparation and pore control of highly mesoporous carbon from defluorinated PTFE. *Carbon.* **2003**, **41***(9)*, 1759–1764.

32. Zhao, J.; et al., Pore structure control of mesoporous carbon as supercapacitor material. *Mater. Lett.* **2007**, **61***(23)*, 4639–4642.

33. Chmiola, J.; et al., Anomalous increase in carbon capacitance at pore sizes less than 1 nanometer. *Sci.* **2006**, **313***(5794)*, 1760–1763.

34. Lin, C.; J.A. Ritter, and B.N. Popov, Correlation of Double-Layer Capacitance with the Pore Structure of Sol-Gel Derived Carbon Xerogels. J. Electrochem. Soc. **1999**, **146***(10)*, 3639–3643.

35. Gogotsi, Y.; et al., Nanoporous carbide-derived carbon with tunable pore size. *Nature Mater.* **2003**, **2***(9)*, 591–594.

36. Ustinov, E.; D. Do, and V. Fenelonov, Pore size distribution analysis of activated carbons: application of density functional theory using nongraphitized carbon black as a reference system. *Carbon.* **2006**. **44***(4)*, 653–663.

37. Han, S.; et al., The effect of silica template structure on the pore structure of mesoporous carbons. *Carbon.* **2003**, **41**(5), 1049–1056.

38. Pérez-Mendoza, M.; et al., Analysis of the microporous texture of a glassy carbon by adsorption measurements and Monte Carlo simulation. Evolution with chemical and physical activation. *Carbon.* **2006**, **44***(4)*, 638–645.

39. Dombrowski, R. J.; D.R. Hyduke, and C.M. Lastoskie, Pore size analysis of activated carbons from argon and nitrogen porosimetry using density functional theory. *Langmuir.* **2000**, **16***(11)*, 5041–5050.

40. Khalili, N. R.; et al., Production of micro-and mesoporous activated carbon from paper mill sludge: I. Effect of zinc chloride activation. Carbon. **2000**, **38***(14)*, 1905–1915.

41. Dandekar, A.; R. Baker, and M. Vannice, Characterization of activated carbon, graphitized carbon fibers and synthetic diamond powder using TPD and DRIFTS. *Carbon.* **1998**, **36***(12)*, 1821–1831.

42. Lastoskie, C.; K.E. Gubbins, and N. Quirke, Pore size distribution analysis of microporous carbons: a density functional theory approach. *J. Phys. Chem.* **1993**, 97*(18)*, 4786–4796.

43. Kakei, K.; et al., Multi-stage micropore filling mechanism of nitrogen on microporous and micrographitic carbons. *J. Chem. Soc. Faraday Trans.* **1990**, 86*(2)*, 371–376.

44. Sing, K. S.; Adsorption methods for the characterization of porous materials. *Adv. Colloid Interface Sci.* **1998**, **76**, 3–11.

45. Barranco, V.; et al., Amorphous carbon nanofibers and their activated carbon nanofibers as supercapacitor electrodes. *J. Phys. Chem. C.* **2010**, 114*(22)*, 10302–10307.

46 Kawabuchi, Y.; et al., Chemical vapor deposition of heterocyclic compounds over active carbon fiber to control its porosity and surface function. *Langmuir.* **1997**, 13*(8)*, 2314–2317.

47. Miura, K.; J. Hayashi, and K. Hashimoto, Production of molecular sieving carbon through carbonization of coal modified by organic additives. *Carbon.* **1991**, 29*(4)*, 653–660.

48. Kawabuchi, Y.; et al., The modification of pore size in activated carbon fibers by chemical vapor deposition and its effects on molecular sieve selectivity. *Carbon.* **1998**, **36***(4)*, 377–382.

49. Verma, S.; and P. Walker, Preparation of carbon molecular sieves by propylene pyrolysis over nickel-impregnated activated carbons. *Carbon.* **1993**. 31*(7)*: p. 1203–1207.

50. Verma, S.; Y. Nakayama, and P. Walker, Effect of temperature on oxygen-argon separation on carbon molecular sieves. Carbon. **1993**, 31*(3)*, 533–534.

51. Chen, Y.; and R. Yang, Preparation of carbon molecular sieve membrane and diffusion of binary mixtures in the membrane. *Ind. Eng. Chem. Res.* **1994**, 33*(12)*, 3146–3153.

52. Rao, M. and S. Sircar, Performance and pore characterization of nanoporous carbon membranes for gas separation. *J. Membrane Sci.* **1996**, 110*(1)*, 109–118.

53. Katsaros, F.; et al., High pressure gas permeability of microporous carbon membranes. *Microporous materials.* **1997**, 8*(3)*, 171–176.

54. Kang, I.; Carbon Nanotube Smart Materials. **2005**, University of Cincinnati.

55. Khan, Z. H.; and Husain, M.; Carbon nanotube and its possible applications. *Indian J. Eng. Mater. Sci.* **2005**. 12*(6)*, 529.

56. Khare, R.; and S. Bose, Carbon nanotube based composites-a review. *J. Minerals Mater. Characterization Eng.* **2005**, 4*(1)*, 31–46.

57. Abuilaiwi, F.A.; et al., Modification and functionalization of multiwalled carbon nanotube (MWCNT) via Fischer esterification. *The Arabian J. Sci. Eng.* **2010**, 35*(1c)*, 37–48.

58. Gupta, S.; and J. Farmer, Multiwalled carbon nanotubes and dispersed nanodiamond novel hybrids: Microscopic structure evolution, physical properties, and radiation resilience. *J. Appl. Phys.* **2011**, 109*(1)*, 014314–014314-15.

59. Upadhyayula, V. K.; and V. Gadhamshetty, Appreciating the role of carbon nanotube composites in preventing biofouling and promoting biofilms on material surfaces in environmental engineering: a review. *Biotechnol. Adv.* **2010**, 28*(6)*: p. 802–816.

60. Saba, J.; et al., Continuous electrodeposition of polypyrrole on carbon nanotube-carbon fiber hybrids as a protective treatment against nanotube dispersion. Carbon. **2012**.

61. Kim, W. D.; et al., Tailoring the carbon nanostructures grown on the surface of Ni–Al bimetallic nanoparticles in the gas phase. *J. Colloid Interface Sci.* **2011**, **362***(2)*: p. 261–266.

62. Schwandt, C.; A. Dimitrov, and D. Fray, The preparation of nano-structured carbon materials by electrolysis of molten lithium chloride at graphite electrodes. *J. Elec. Chem.* **2010**, **647***(2)*, 150–158.

63. Gao, C.; et al., The new age of carbon nanotubes: An updated review of functionalized carbon nanotubes in electrochemical sensors. *Nanoscale.* **2012**, *4(6)*, 1948–1963.

64. Ben-Valid, S.; et al., Spectroscopic and electrochemical study of hybrids containing conductive polymers and carbon nanotubes. *Carbon.* **2010**, *48(10)*, 2773–2781.

65. Vecitis, C. D.; G. Gao, and H. Liu, Electrochemical carbon nanotube filter for adsorption, desorption, and oxidation of aqueous dyes and anions. *J. Phys. Chem. C.* **2011**, *115(9)*, 3621–3629.

66. Jagannathan, S.; et al., Structure and electrochemical properties of activated polyacrylonitrile based carbon fibers containing carbon nanotubes. *J. Power Sources.* **2008**, *185(2)*, 676–684.

67. Zhu, Y.; et al., Carbon-based supercapacitors produced by activation of graphene. *Sci.* **2011**, *332(6037)*, 1537–1541.

68. Obreja, V. V.; On the performance of supercapacitors with electrodes based on carbon nanotubes and carbon activated material—a review. *Physica E: Low-dimensional Systems and Nanostructures.* **2008**, *40(7)*, 2596–2605.

69. Wang, L.; and Yang, R. T.; Hydrogen storage on carbon-based adsorbents and storage at ambient temperature by hydrogen spillover. *Catalysis Rev.: Sci. Eng.* **2010**, *52(4)*, 411–461.

70. Dillon, A.; et al., Carbon nanotube materials for hydrogen storage. Proceedings of the 1997 DOE/NREL Hydrogen Program Review, **1997**, 237 p.

71. Ströbel, R.; et al., Hydrogen storage by carbon materials. *J. Power Sources.* **2006**, *159(2)*, 781–801.

72. V. K. Gupta & T. A. Saleh. Sorption of pollutants by porous carbon, carbon nanotubes and fullerene. *An Overview, Environ. Sci. Pollut. Res.* **2013**, *20*, 2828–2843.

73. V. K. Gupta, et al., Adsorptive removal of dyes from aqueous solution onto carbon nanotubes: a review. *Adv. Colloid Interface Sci.* **2013**, *193–194*, 24–34.

74. Yang, K.; et al., Competitive sorption of pyrene, phenanthrene, and naphthalene on multiwalled carbon nanotubes. *Environ. Science Technol.* **2006**, *40(18)*, 5804–5810.

75. Zhang, H.; et al., Synthesis of a novel composite imprinted material based on multiwalled carbon nanotubes as a selective melamine absorbent. *J. Agricult. Food Chem.* **2011**. *59(4)*: p. 1063–1071.

76. Yan, L.; et al., Characterization of magnetic guar gum-grafted carbon nanotubes and the adsorption of the dyes. *Carbohydrate Polym.* **2012**, *87(3)*, 1919–1924.

77. Wu, C. H.; Adsorption of reactive dye onto carbon nanotubes: equilibrium, kinetics and thermodynamics. *J. Hazardous Mater.* **2007**, *144(1–2)*, 93–100.

78. Vadi, M. and E. Ghaseminejhad, comparative study of isotherms adsorption of oleic acid by activated carbon and multi-wall carbon nanotube. *Oriental J. Chem.* **2011**, *27(3)*, 973.

79. Vadi, M. and N. Moradi, Study of Adsorption Isotherms of Acetamide and Propionamide on Carbon Nanotube. *Oriental J. Chem.* **2011**, *27(4)*, 1491.

80. Mishra, A.K.; T. Arockiadoss, and S. Ramaprabhu, Study of removal of azo dye by functionalized multi walled carbon nanotubes. *Chem. Eng. J.* **2010**, *162(3)*, 1026–1034.

81. Madrakian, T.; et al., Removal of some cationic dyes from aqueous solutions using magnetic-modified multi-walled carbon nanotubes. *J. Hazardous Mater.* **2011**, *196*, 109–114.

82. Kuo, C. Y.; C.H. Wu, and J.Y. Wu, Adsorption of direct dyes from aqueous solutions by carbon nanotubes: Determination of equilibrium, kinetics and thermodynamics parameters. *J. Colloid Interface Sci.* **2008**, *327(2)*, 308–315.

83. Gong, J. L.; et al., Removal of cationic dyes from aqueous solution using magnetic multi-wall carbon nanotube nanocomposite as adsorbent. *J. Hazardous Mater.* **2009**, *164(2)*, 1517–1522.

84. Chang, P.R.; et al., Characterization of magnetic soluble starch-functionalized carbon nanotubes and its application for the adsorption of the dyes. *J. Hazardous Mater.* **2011**, *186(2)*, 2144–2150.

85. Chatterjee, S.; M.W. Lee, and S.H. Woo, Adsorption of Congo red by chitosan hydrogel beads impregnated with carbon nanotubes. *Bioresource Technol.* **2010**, *101(6)*, 1800–1806.

86. Chen, Z.; et al., Adsorption behavior of epirubicin hydrochloride on carboxylated carbon nanotubes. *Int. J. Pharmaceutics.* **2011**. *405(1)*, 153–161.

87. Ai, L.; et al., Removal of methylene blue from aqueous solution with magnetite loaded multi-wall carbon nanotube: kinetic, isotherm and mechanism analysis. *J. Hazardous Mater.* **2011**, *198*, 282–290.

88. Chronakis, I.S.; Novel nanocomposites and nanoceramics based on polymer nanofibers using electrospinning process—a review. *J. Mater. Proc. Technol.* **2005**, *167(2)*, 283–293.

89. Inagaki, M.; Y. Yang, and F. Kang, Carbon nanofibers prepared via electrospinning. *Adv. Mater.* **2012**, *24(19)*, 2547–2566.

90. Im, J. S.; et al., The study of controlling pore size on electrospun carbon nanofibers for hydrogen adsorption. *J. Colloid Interface Sci.* **2008**, *318(1)*, 42–49.

91. De Jong, K. P.; and Geus, J. W.; Carbon nanofibers: catalytic synthesis and applications. *Catalysis Rev.* **2000**, *42(4)*, 481–510.

92. Huang, J.; Liu, Y.; and You, T.; Carbon nanofiber based electrochemical biosensors: a review. *Anal. Methods.* **2010**, *2(3)*, 202–211.

93. Yusof, N.; and Ismail, A.; Post spinning and pyrolysis processes of polyacrylonitrile (PAN)-based carbon fiber and activated carbon fiber: A review. *J. Anal. Appl. Pyrolysis* **2012**, *93*, 1–13.

94. Sullivan, P.; et al. Physical and chemical properties of PAN-derived electrospun activated carbon nanofibers and their potential for use as an adsorbent for toxic industrial chemicals. *Adsorption.* **2012**, *18(3–4)*, 265–274.

95. Tavanai, H.; Jalili, R.; and Morshed, M.; Effects of fiber diameter and CO_2 activation temperature on the pore characteristics of polyacrylonitrile based activated carbon nanofibers. *Surface Interface Analy.* **2009**, *41(10)*, 814–819.

96. Wang, G.; et al., Activated carbon nanofiber webs made by electrospinning for capacitive deionization. *Electrochimica Acta.* **2012**, *69*, 65–70.

97. Ra, E. J.; et al., Ultramicropore formation in PAN/camphor-based carbon nanofiber paper. *Chem. Commun.* **2010**, *46(8)*, 1320–1322.

98. Liu, W.; and S. Adanur, Properties of electrospun polyacrylonitrile membranes and chemically-activated carbon nanofibers. *Text. Res. J.* **2010**, *80(2)*, 124–134.

99. Korovchenko, P.; Renken, A.; and Kiwi-Minsker, L.; Microwave plasma assisted preparation of Pd-nanoparticles with controlled dispersion on woven activated carbon fibres. *Catalysis Today.* **2005**, *102*, 133–141.

100. Jung, K. H.; and Ferraris, J. P.; Preparation and electrochemical properties of carbon nanofibers derived from polybenzimidazole/polyimide precursor blends. *Carbon.* **2012**.

101. Im, J. S.; Park, S. J.; and Lee, Y. S.; Preparation and characteristics of electrospun activated carbon materials having meso-and macropores. *J. Colloid Interface Sci.* **2007**. *314(1):* p. 32–37.

102. Esrafilzadeh, D.; M. Morshed, and H. Tavanai, An investigation on the stabilization of special polyacrylonitrile nanofibers as carbon or activated carbon nanofiber precursor. *Synthetic Metals.* **2009**, *159(3)*, 267–272.

103. Hung, C. M.; Activity of Cu-activated carbon fiber catalyst in wet oxidation of ammonia solution. *J. Hazardous Mater.* **2009**, *166(2)*, 1314–1320.

104. Koslow, E. E.; Carbon or activated carbon nanofibers. **2007**, Google Patents.

105. Lee, J. W.; et al., Heterogeneous adsorption of activated carbon nanofibers synthesized by electrospinning polyacrylonitrile solution. *J. Nanosci. Nanotechnol.* **2006**, *6(11)*, 3577–3582.

106. Oh, G. Y.; et al., Preparation of the novel manganese-embedded PAN-based activated carbon nanofibers by electrospinning and their toluene adsorption. *J. Anal. Appl. Pyrolysis.* **2008**, *81(2)*, 211–217.

107. Zussman, E.; et al., Mechanical and structural characterization of electrospun PAN-derived carbon nanofibers. Carbon. **2005**, *43(10)*, 2175–2185.

108. Kim, C.; Electrochemical characterization of electrospun activated carbon nanofibres as an electrode in supercapacitors. *J. Power Sources.* **2005**, *142(1)*, 382–388.

109. Jung, M. J.; et al., Influence of the textual properties of activated carbon nanofibers on the performance of electric double-layer capacitors. *J. Ind. Eng. Chem.* **2013**.

110. Fan, Z.; et al., Asymmetric supercapacitors based on graphene/MnO_2 and activated carbon nanofiber electrodes with high power and energy density. *Adv. Funct. Mater.* **2011**, *21(12)*, 2366–2375.

111. Jeong, E.; M.J. Jung, and Y.S. Lee, Role of fluorination in improvement of the electrochemical properties of activated carbon nanofiber electrodes. *J. Fluorine Chem.* **2013**.

112. Seo, M. K.; and S.J. Park, Electrochemical characteristics of activated carbon nanofiber electrodes for supercapacitors. *Mater. Sci. Eng.: B.* **2009**. *164(2):* p. 106–111.

113. Endo, M.; et al., High power electric double layer capacitor (EDLC's); from operating principle to pore size control in advanced activated carbons. *Carbon Sci.* **2001**, *1(3&4)*, 117–128.

114. Ji, L.; and Zhang, X.; Generation of activated carbon nanofibers from electrospun polyacrylonitrile-zinc chloride composites for use as anodes in lithium-ion batteries. *Electrochem. Commun.* **2009**, *11(3)*, 684–687.

115. Karra, U.; et al., Power generation and organics removal from wastewater using activated carbon nanofiber (ACNF) microbial fuel cells (MFCs). *Int. J. Hydrogen Ener.* **2012**.

116. Oh, G. Y.; et al., Adsorption of toluene on carbon nanofibers prepared by electrospinning. *Sci. Total Environ.* **2008**, *393(2)*, 341–347.

117. Lee, K.J.; et al. Activated carbon nanofiber produced from electrospun polyacrylonitrile nanofiber as a highly efficient formaldehyde adsorbent. *Carbon.* **2010**, *48(15)*, 4248–4255.

118. Katepalli, H.; et al. Synthesis of hierarchical fabrics by electrospinning of PAN nanofibers on activated carbon microfibers for environmental remediation applications. *Chem. Eng. J.* **2011**, *171(3)*, 1194–1200.

119. Gaur, V.; Sharma, A.; and Verma, N.; Preparation and characterization of ACF for the adsorption of BTX and SO$_2$. *Chem. Eng. Proc.: Process Int.* **2006**. **45***(1)*: p. 1–13.

120. Cheng, K. K.; T.C. Hsu, and L.H. Kao, A microscopic view of chemically activated amorphous carbon nanofibers prepared from core/sheath melt-spinning of phenol formaldehyde-based polymer blends. *J. Mater. Sci.* **2011**, *46(11)*, 3914–3922.

121. Zhijun, et al. A molecular simulation probing of structure and interaction for supramolecular sodium dodecyl sulfate/s-w carbon nanotube assemblies. *Nano Lett.* **2009**.

122. Ong, et al. Molecular dynamics simulation of thermal boundary conductance between carbon nanotubes and sio$_2$. *Phys. Rev. B.* **2010**, *81*.

123. Yang, et al., Preparation and properties of phenolic resin-based activated carbon spheres with controlled pore size distribution, Carbon, **2002**, **40***(5)*: p. 911–916.

124. Liu, et al., Carbon nanotube based artificial water channel protein: membrane perturbation and water transportation. *Nano Lett.* **2009**, *9(4)*, 1386–1394.

125. Zhang, et al., Interfacial Characteristics of Carbon Nanotube-Polyethylene Composites Using Molecular Dynamics Simulations, ISRN Materials Science, Article ID 145042, **2011**.

126. L. Mangun, et al., Surface chemistry, pore sizes and adsorption properties of activated carbon fibers and precursors treated with ammonia, Carbon, **2001**, *139(11)*, 1809–1820.

127. Irle, et al., Milestones in molecular dynamics simulations of single-walled carbon nanotube formation: a brief critical review. *Nano Res.* **2009**, *2(8)*, 755–767.

128. Guo-zhuo, et al., Regulation of pore size distribution in coal-based activated carbon. *New Carbon Mater.* **2009**, *24(7820)*, 8827–1007.

129. Zang, et al. A comparative study of Young's modulus of single-walled carbon nanotube by CPMD, MD and first principle simulations. *Comput. Mater. Sci.* **2009**, *46(4)*, 621–625.

130. Frankland, et al. Molecular simulation of the influence of chemical cross-links on the shear strength of carbon nanotube-polymer interfaces. *J. Phys. Chem.* B, **2002**, *106(2)*, 3046–3048.

131. Frankland, et al. Simulation for separation of hydrogen and carbon monoxide by adsorption on single-walled carbon nanotubes. *Fluid Phase Equilibria.* **2002**, *194–197(10)*, 297–307.

132. Han, et al., Molecular dynamics simulations of the elastic properties of polymer/carbon nanotube composites. *Comput. Mater. Sci.* **2007**, **39***(8)*, 315–323.

133. Okhovat, et al. Pore Size Distribution Analysis of Coal-Based Activated Carbons: Investigating the Effects of Activating Agent and Chemical Ratio, ISRN Chemical Engineering; **2012**.

134. Lehtinen, et al., Effects of ion bombardment on a two-dimensional target: Atomistic simulations of graphene irradiation. *Phys. Rev. B,* **2010**, 81.

135. Shigeo Maruyama, A Molecular Dynamics Simulation Of Heat Conduction Of A Finite Length Single-Walled Carbon Nanotube, Microscale Thermophysical Engineering; **2003**, *7,* 41–50.

136. Williams, et al., Monte carlo simulations of H$_2$ physisorption in finite-diameter carbon nanotube ropes. *Chem. Phys. Lett.* **2000**, *320(6),* 352–358.

137. Diao, et al., Molecular dynamics simulations of carbon nanotube/silicon interfacial thermal conductance. *J. Chem. Phys.* **2008**, 128.

138. Noel, et al. On the use of symmetry in the ab initio quantum mechanical simulation of nanotubes and related materials. *J. Comput Chem.* **2010**, *31,* 855–862.

139. Nicholls, D.; et al., *Water Transport Through (7,7) Carbon Nanotubes of Different Lengths using Molecular Dynamics, Microfluidics and Nanofluidics.* **2012**, *1–4,* 257–264.

140. Ribas, A.; et al. Nanotube nucleation versus carbon-catalyst adhesion–Probed by molecular dynamics simulations. *J. Chem. Phys.* **2009**, 131.

141. Sanz-Navarro, F.; et al., Molecular dynamics simulations of metal clusters supported on fishbone carbon nanofibers. *J. Phys. Chem. C.* **2010**, *114,* 3522–3530.

142. Shibuta, et al. Bond-Order Potential for Transition Metal Carbide Cluster for the Growth Simulation of a Single-Walled Carbon Nanotube. Department of Materials Engineering, The University of Tokyo; **2003**.

143. Mao, et al. Molecular dynamics simulations of the filling and decorating of carbon nanotubules. *Nanotechnology.* **1999**, *10,* 273–277.

144. Mao, et al. Molecular simulation study of ch$_4$/h$_2$ mixture separations using metal organic framework membranes and composites. *J. Phys. Chem. C.* **2010**, *114,* 13047–13054.

145. Thomas, A.; et al. Pressure-driven Water Flow through Carbon Nanotubes: Insights from Molecular Dynamics Simulation, Carnegie Mellon University, USA, Department of Mechanical Engineering; **2009**.

CHAPTER 6

REINFORCEMENT OF POLYMER NANOCOMPOSITES: VARIETY OF STRUCTURAL FORMS AND APPLICATIONS

G. V. KOZLOV, YU. G. YANOVSKII, and G. E. ZAIKOV

CONTENTS

6.1   INTRODUCTION

The experimental analysis of particulate-filled nanocomposites butadiene—styrene rubber/fullerene-containing mineral (nanoshungite) was fulfilled with the aid of force-atomic microscopy, nanoindentation methods, and computer treatment. The theoretical analysis was carried out within the frameworks of fractal analysis. It has been shown that interfacial regions in the aforementioned nanocomposites are the same reinforcing element as nanofiller. The conditions of the transition from nano to microsystems were discussed. The fractal analysis of nanoshungite particles aggregation in polymer matrix was performed. It has been shown that reinforcement of the studied nanocomposites is a true nanoeffect.

The modern methods of experimental and theoretical analysis of polymer materials structure and properties allow not only to confirm earlier propounded hypotheses, but also to obtain principally new results. Let us consider some important problems of particulate-filled polymer nanocomposites, the solution of which allows to advance substantially in these materials' properties understanding and prediction.

Polymer nanocomposites multicomponentness (multiphaseness) requires their structural components to be quantitative characteristics determination. In this aspect, interfacial regions play a particular role, as it has been shown earlier, that they are the same reinforcing element in elastomeric nanocomposites as nanofiller actually [1]. Therefore, the knowledge of interfacial layer dimensional characteristics is necessary for quantitative determination of one of the most important parameters of polymer composites, in general,—their reinforcement degree [2, 3].

The aggregation of the initial nanofiller powder particles in small or large particle aggregates always occurs in the course of technological process of making particulate-filled polymer composites in general [4] and elastomeric nanocomposites in particular [5]. The aggregation process explains composites (nanocomposites) macroscopic properties [2–4]. For nanocomposites, nanofiller aggregation process gains special significance, as its intensity can be the one, that nanofiller particle aggregates size exceed 100 nm of—the value, which is assumed (though conditionally enough [6]) as an upper dimensional limit for nanoparticle.

In other words, the aggregation process can result to the situation when primordially supposed nanocomposite ceases to be one. Therefore, at present several methods exist, which allow to suppress nanoparticles aggregation process [5, 7]. This also assumes the necessity of the nanoparticles aggregation process as quantitative analysis. It is well-known [1, 2], that in particulate-filled elastomeric nanocomposites (rubbers) nanofiller particles form linear spatial structures ("chains"). At the same time in polymer composites, filled with disperse microparticles (microcomposites), particles (aggregates of particles) of filler form a fractal network, which defines polymer matrix structure (analog of fractal lattice in computer simulation) [4]. This results to different mechanisms of polymer matrix structure formation in micro- and nanocomposites. If, in the first filler particles (aggregates of particles), fractal network availability results to "disturbance" of polymer matrix structure, which is expressed in the increase of its fractal dimension d_f [4], then in case of polymer nanocomposites at nanofiller contents change, the value d_f is not changed and equal to matrix polymer structure fractal dimension [3]. As it has been expected, the change of the composites of the indicated classes structure formation mechanism defines their properties, in particular, reinforcement degree [9, 11, 12]. Therefore nanofiller structure fractality strict proof and its dimension determination are necessary.

As it is known [13, 14], the scale effects are often found at the study of different materials mechanical properties. The dependence of failure stress on grain size for metals (Holl-Petsch formula) [15] or of effective filling degree on filler particles size in case of polymer composites [16] are examples of such effect. The strong dependence of elasticity modulus on nanofiller particles diameter is observed for particulate-filled elastomeric nanocomposites [5]. Therefore, it is necessary to elucidate the physical grounds of nano- and micromechanical behavior scale effect for polymer nanocomposites.

At present a disperse material wide list is known, which is able to strengthen elastomeric polymer materials [5]. These materials are very diverse on their surface chemical constitution, but the small size of particles is a common feature for them. Based on the observation the hypothesis was offered that any solid material would strengthen the rubber at the condition, which it was in a very-dispersed state and could be dispersed in polymer matrix. Edwards [5] points out that filler particles small size is necessary and, probably, the main requirement for reinforcement effect realization in rubbers. Using modern terminology, the nanofiller particles, for which their aggregation process is suppressed as far as possible, would be the most effective ones for rubbers reinforcement [3, 12]. Therefore, the theoretical analysis of a nanofiller particles' size influence on polymer nanocomposites reinforcement is necessary.

Based on the aforementioned discussion, the purpose of this work is to provide the solution of the aforementioned paramount problems with the help of modern experimental and theoretical techniques on the example of particulate-filled butadiene—styrene rubber.

6.2 EXPERIMENTAL

The industrially made butadiene—styrene rubber of mark SKS-30, which contains 7.0–12.3 per cent cis and 71.8–72.0 per cent trans-bonds, with density of 920–930 kg/m³ was used as matrix polymer. This rubber is fully amorphous one. Fullerene-containing mineral shungite of Zazhoginsk's deposit consists of ~30 per cent globular amorphous metastable carbon and ~70 per cent high-disperse silicate particles. Besides, industrially made technical carbon of mark no 220 was used as nanofiller.

The average size of technical carbon, nano, and microshugite particles makes up 20, 40, and 200 nm, respectively. The indicated filler content is equal to 37 mass per cent. Nano and microdimensional disperse shungite particles were prepared from industrially output material by the original technology processing. The size and polydispersity analysis of the received in milling process shungite particles was monitored with the aid of analytical disk centrifuge (CPS Instruments, Inc., USA), allowing to determine with high precision size and distribution by the sizes within the range from 2 nm to 50 mcm.

Nanostructure was studied on atomic-forced microscopes Nano-DST (Pacific Nanotechnology, USA) and Easy Scan DFM (Nanosurf, Switzerland) by semi-contact method in the force modulation regime. Atomic-force microscopy results were processed with the help of specialized software package scanning probe image processor (SPIP). The SPIP is a powerful program package for processing images, obtained on SPM, AFM, STM, scanning electron microscopes, transmission electron microscopes, interferometers, confocal microscopes, profilometers, optical microscopes, and so on. The given package possesses the whole function number, which is necessary at images precise analysis, in a number of which the following ones are included:

-   the possibility of three-dimensional reflecting objects obtaining, distortions automatized leveling, including Z-error mistakes removal for examination of separate elements, and so on;
-   quantitative analysis of particles or grains, >40 parameters can be calculated for each found particle or pore: area, perimeter, mean diameter, the ratio of linear sizes of grain width to its height distance between grains, coordinates of grain center of mass can be presented in a diagram form or in a histogram form.

The tests on elastomeric nanocomposites nanomechanical properties were carried out by a nanointentation method [17] on apparatus Nano Test 600 (Micro Materials, Great Britain) in loades wide range from 0.01 mN to 2.0 mN. Sample indentation was conducted in 10 points with interval of 30 mcm. The load was increased with constant rate up to the greatest given load reaching (for the rate 0.05 mN/s^{-1} mN). The indentation rate was changed in conformity with the greatest load value counting, that loading cycle should take 20 s.

The unloading was conducted with the same rate as loading. In the given experiment the "Berkovich indentor" was used with the angle at the top of 65.3° and rounding radius of 200 nm. Indentations were carried out in the checked load regime with preload of 0.001 mN. For elasticity modulus calculation the obtained result in the experiment by nanoindentation course dependences of load on indentation depth (strain) in 10 points for each sample at loads of 0.01, 0.02, 0.03, 0.05, 0.10, 0.50, 1.0, and 2.0 mN were processed according to Oliver-Pharr method [18].

6.3   RESULTS AND DISCUSSION

In Figure 6.1, the obtained data according to the original methodics results of elasticity moduli calculation for nanocomposite butadiene—styrene rubber/nanoshungite components (matrix, nanofiller particle, and interfacial layers), received in interpolation process of nanoindentation data, are presented. The processed in SPIP polymer nanocomposite image with shungite nanoparticles allows experimental determination of interfacial layer thickness l_{if}, which is presented in Figure 6.1 as steps on elastomeric matrix-nanofiller boundary.

The measurements of 34 such steps' (interfacial layers) width on the processed SPIP images of interfacial layer's various section gave the mean experimental value $l_{if} = 8.7$ nm. Besides, nanoindentation results (Figure 6.1, figures on the right) showed that interfacial layers elasticity modulus was only 23–45 per cent lower than nanofiller elasticity modulus, but higher than the corresponding parameter of polymer matrix in 6.0–8.5 times. These experimental data confirm that for the studied nanocomposite interfacial layer is a reinforcing element to the same extent as nanofiller actually [1, 3, 12].

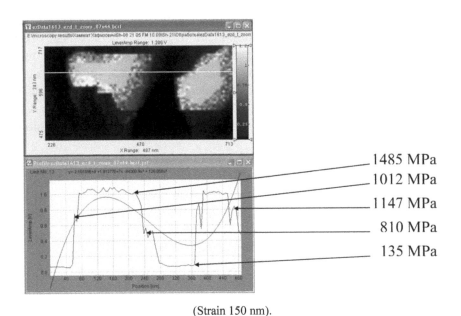

1485 MPa

1012 MPa

1147 MPa

810 MPa

135 MPa

(Strain 150 nm).

FIGURE 6.1   The processed SPIP image of nanocomposite butadiene—styrene rubber/ nanoshungite, obtained by force modulation method, and mechanical characteristics of structural components.

Source: According to the Data of Nanoindentation (Strain 150 nm).

Let us fulfill the value l_{if} theoretical estimation according to the two methods and compare these results with the ones obtained experimentally. The first method simulates interfacial layer in polymer composites as a result of interaction of two fractals—polymer matrix and nanofiller surface [19, 20]. In this case, there is a sole linear scale l, which defines these fractals interpenetration distance [21]. As nanofiller elasticity modulus is essentially higher than the corresponding parameter for rubber (in the considered case—in 11 times, see Figure 6.1), then the indicated interaction reduces to nanofiller indentation in polymer matrix and then $l = l_{if}$. In this case, it can be written as in what follows [21]:

$$l_{if} \approx a \left(\frac{R_p}{a} \right)^{2(d-d_{surf})/d}, \tag{6.1}$$

where a is a lower linear scale of fractal behavior, which is accepted for polymers as equal to statistical segment length l_{st} [22], R_p is a nanofiller particle (more precisely, particle aggregates) radius, which for nanoshungite is equal to ~84 nm [23], d is dimension of Euclidean space, in which fractal is considered (here $d = 3$), d_{surf} is fractal dimension of nanofiller particle aggregate surface.

The value l_{st} is determined as follows [24]:

$$l_{st} = l_0 C_\infty, \tag{6.2}$$

where l_0 is the main chain skeletal bond length, which is equal to 0.154 nm for both blocks of butadiene—styrene rubber [25], C is characteristic ratio, which is a polymer chain statistical flexibility indicator [26], and is determined with the help of Eq. (6.3) [22]:

$$T_g = 129 \left(\frac{S}{C_\infty} \right)^{1/2}, \tag{6.3}$$

where T_g is glass transition temperature, equal to 217 K for butadiene—styrene rubber [3], S is macromolecule cross-sectional area, determined for the mentioned rubber according to the additivity rule from the following considerations. As it is known [27], the macromolecule diameter quadrate values are equal for polybutadiene −20.7 Å^2 and for polystyrene −69.8 Å^2. Having calculated cross-sectional area of macromolecule, simulated as a cylinder, for the indicated polymers according to the known geometrical formulas, let us obtain 16.2 and 54.8 Å^2, respectively. Further, accepting S as the average value of the adduced above areas, let us obtain for butadiene—styrene rubber $S = 35.5$ Å^2. Then, according to Eq. (6.3) at the indicated values T_g and S let us obtain $C = 12.5$ and according to Eq. (6.2)—$l_{st} = 1.932$ nm.

The fractal dimension of nanofiller surface d_{surf} was determined with the help of the equation [3]:

$$S_u = 410 R_p^{d_{surf}-d}, \tag{6.4}$$

where S_u is nanoshungite particles specific surface, calculated as in what follows [28]:

$$S_u = \frac{3}{\rho_n R_p},$$

(6.5)

where ρ_n is the nanofiller particle aggregate density, determined according to the formula [3]:

$$\rho_n = 0.188(R_p)^{1/3}.$$

(6.6)

The calculation according to Eqs. (6.4)–(6.6) gives d_{surf} = 2.44. Further, based on the calculation by the indicated mode parameters, let us obtain the theoretical value of interfacial layer thickness l_f^T = 7.8 nm, from Eq. (6.1). This value is close enough to the obtained one experimentally (their discrepancy makes up ~10%).

The second method of value l_{if}^T estimation consists of the two following equations [3, 29]:

$$\phi_{if} = \phi_n(d_{surf} - 2)$$

(6.7)

and

$$\phi_{if} = \phi_n\left[\left(\frac{R_p + l_{if}^T}{R_p}\right)^3 - 1\right],$$

(6.8)

where ϕ_{if} and ϕ_n are relative volume fractions of interfacial regions and nanofiller, accordingly.

The combination of the indicated equations allows to receive the following formula for l_{if}^T calculation:

$$l_{if}^T = R_p\left[(d_{surf} - 1)^{1/3} - 1\right].$$

(6.9)

The calculation according to Eq. (6.9) gives l_{if}^T = 10.8 nm for the considered nanocomposite, which also corresponds well enough to the experiment (in this case discrepancy between l_{if} and l_{if}^T makes up ~19%).

In conclusion, let us note the important experimental observation, which follows from the processed data by program SPIP results of the studied nanocomposite surface scan (Figure 6.1). As one can see, at one nanoshungite particle surface from one to three (in average—two) steps can be observed, structurally identified as interfacial layers. It is significant that these steps width (or l_{if}) is approximately equal to the first (the closest to nanoparticle surface) step width. Therefore, the indicated observation supposes that in elastomeric nanocomposites an average of two interfacial layers are formed: the first is at the expense of nanofiller particle surface with elastomeric matrix interaction, as a result of which molecular mobility in this layer is frozen and its state is glassy-like one, and the second—at the expense of glassy interfacial layer with elastomeric polymer matrix interaction. The most important question from the practical point of view is whether one interfacial layer or both serve as nanocomposite reinforcing element. Let us fulfill the following quantitative estimation for this question. The reinforcement degree (E_n/E_m) of polymer nanocomposites is given by the equation in what follows [3]:

$$\frac{E_n}{E_m} = 1 + 11\left(\phi_n + \phi_{if}\right)^{1.7}, \tag{6.10}$$

where E_n and E_m are elasticity moduli of nanocomposite and matrix polymer, respectively ($E_m = 1.82$ MPa [3]).

According to Eq. (6.7) the sum ($\varphi_n + \varphi_{if}$) is equal to:

$$\phi_n + \phi_{if} = \phi_n\left(d_{surf} - 1\right), \tag{6.11}$$

if one interfacial layer (the closest to nanoshungite surface) is a reinforcing element and

$$\phi_n + 2\phi_{if} = \phi_n\left(2d_{surf} - 3\right), \tag{6.12}$$

if both interfacial layers are reinforcing elements.

The value φ_n is determined according to the equation in what follows [30]:

$$\phi_n = \frac{W_n}{\rho_n}, \tag{6.13}$$

where W_n is nanofiller mass content, ρ_n is its density, determined according to Eq. (6.6).

The calculation according to Eqs. (6.11) and (6.12) gave the following E_n/E_m values: 4.60 and 6.65, respectively. Since the experimental value $E_n/E_m = 6.10$ is closer to the value, calculated according to Eq. (6.12), then both interfacial layers are the reinforcing elements for the studied nanocomposites. Therefore, the coefficient 2 should be introduced in the equations for value l_{if} determination (e.g. in Eq. (6.1)) in case of nanocomposites with elastomeric matrix. Let us consider that Eq. (6.1) in its initial form was obtained as a relationship with proportionality sign, i.e., without fixed proportionality coefficient [21].

Thus, the aforementioned used nanoscopic methodics allow estimating both interfacial layer and structural special features in polymer nanocomposites and its sizes and properties. For the first time it has been shown that two consecutive interfacial layers are formed in elastomeric particulate-filled nanocomposites, which are reinforcing elements for the indicated nanocomposites. The proposed theoretical methodics of interfacial layer thickness estimation, elaborated within the frameworks of fractal analysis, give well enough correspondence to the experiment.

For theoretical treatment of nanofiller particle aggregate growth processes, and final sizes, traditional irreversible aggregation models are inapplicable, because it is obvious that in nanocomposite aggregates a large number of simultaneous growth takes place. Therefore, the model of multiple growth, offered in Ref. [6], was used for nanofiller aggregation description.

In Figure 6.2, the images of the studied nanocomposites, obtained in the force modulation regime, and corresponding nanoparticle aggregates fractal dimension d_f distributions are adduced. As it follows from the adduced values d_f^{ag} ($d_f^{ag} = 2.40$–2.48), nanofiller particle aggregates in the studied nanocomposites are formed by a mechanism particle-cluster (P–Cl), i.e., they are Witten—Sander clusters [32]. The variant A, was chosen which according to mobile particles are added to the lattice, consisting of a large number of "seeds" with density of c_0 at beginning of simulation . Such model generates the structures, which have fractal geometry on short length scales with value $d_f \approx 2.5$ (see Figure 6.2) and homogeneous structure on large length scales. A relatively high particle concentration c is required in the model for uninterrupted network formation [31].

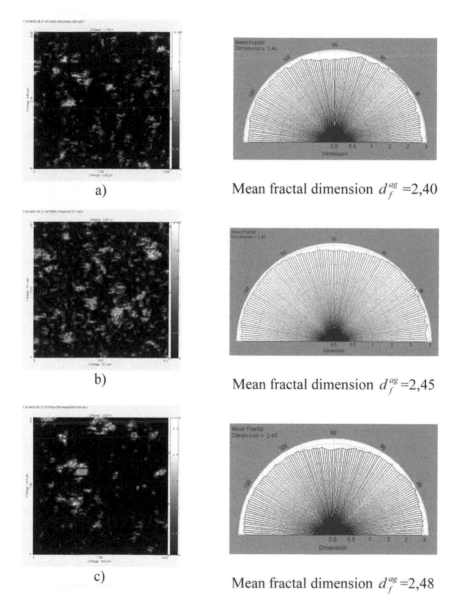

a)

Mean fractal dimension $d_f^{ag} = 2,40$

b)

Mean fractal dimension $d_f^{ag} = 2,45$

c)

Mean fractal dimension $d_f^{ag} = 2,48$

FIGURE 6.2 The images, obtained in the force modulation regime, for nanocomposites, filled with technical carbon (a) nanoshungite, (b) microshungite, and (c) corresponding to them fractal dimensions d_f^{ag}.

In case of "seeds" high concentration c_0 for the variant A, the following relationship was obtained [31]:

$$R_{max}^{d_f^{ag}} = N = c / c_0,$$ (6.14)

where R_{max} is nanoparticles' cluster (aggregate) greatest radius, N is nanoparticles' number per one aggregate, c is nanoparticles' concentration, c_0 is "seeds" number, which is equal to nanoparticles' clusters (aggregates) number.

The value N can be estimated according to the equation in what follows [8]:

$$2R_{max} = \left(\frac{S_n N}{\pi\eta}\right)^{1/2},$$ (6.15)

where S_n is cross-sectional area of nanoparticles, of which an aggregate consists, η is a packing coefficient, which is equal to 0.74 [28].

The experimentally obtained nanoparticle aggregate diameter $2R_{ag}$ was accepted as $2R_{max}$ (Table 6.1) and the value S_n was also calculated according to the experimental values of nanoparticle radius r_n (Table 6.1). In Table 6.1, the values N for the studied nanofillers, obtained according to the indicated method, were adduced. It is significant that the value N is maximum for nanoshungite despite larger values r_n in comparison with technical carbon.

Furthermore, Eq. (6.14) allows to estimate the greatest radius R_{max}^T of nanoparticle aggregate within the frameworks of the aggregation model [31]. These values R_{max}^T are adduced in Table 6.1, from which their reduction in a sequence of technical carbon—nanoshungite—microshungite, that fully contradicts to the experimental data, i.e., to R_{ag} change (Table 6.1). However, we must not neglect the fact that Eq. (6.14) was obtained within the frameworks of computer simulation, where the initial aggregating particle sizes are the same in all cases [31]. For real nanocomposites the values r_n can be distinguished essentially (Table 6.1). It is expected that the value R_{ag} or R_{max}^T will be the higher, the larger is the radius of nanoparticles, forming aggregate, i.e., r_n. Then, the theoretical value of nanofiller particle cluster (aggregate) radius R_{ag}^T can be determined as follows:

$$R_{ag}^T = k_n r_n N^{1/d_f^{ag}} , \qquad (6.16)$$

where k_n is the proportionality coefficient, in the present work accepted empirically equal to 0.9.

TABLE 6.1 The parameters of irreversible aggregation model of nanofiller particles aggregates growth

Nanofiller	R_{ag} (nm)	r_n (nm)	N	R_{max}^T (nm)	R_g^T (nm)	R_c (nm)
Technical carbon	34.6	10	35.4	34.7	34.7	33.9
Nanoshungite	83.6	20	51.8	45.0	90.0	71.0
Microshungite	117.1	100	4.1	15.8	158.0	255.0

The comparison of experimental R_{ag} and calculated, according to the Eq. (6.16), R_{ag}^T values of the studied nanofiller particle aggregates radius shows their good correspondence (the average discrepancy of R_{ag} and R_g^T makes up to 11.4%). Therefore, the theoretical model [31] gives a good correspondence to the experiment only in case of consideration of aggregating particles real characteristics and, in the first place, their size.

Let us consider two more important aspects of nanofiller particles aggregation within the frameworks of the model [31]. Some features of the indicated process are defined by nanoparticle diffusion at nanocomposite processing. Specifically, length scale, connected with diffusible nanoparticle, is correlation length ξ of diffusion. By definition, the growth phenomena in sites, remote more than ξ, are statistically independent. Such definition allows to connect the value ξ with the mean distance between nanofiller particle aggregates L_n. The value ξ can be calculated according to the equation as in what follows [31]:

$$\xi^2 \approx \tilde{n}^{-1} R_{ag}^{d_f^{ag} - d + 2} , \qquad (6.17)$$

where c is nanoparticles concentration, which should be accepted equal to nanofiller volume contents φ_n, which is calculated according to Eqs. (6.6) and (6.13).

The values r_n and R_{ag} were obtained experimentally (see histogram of Figure 6.3). In Figure 6.4 the relation between L_n and ξ is adduced, which, as it is expected, proves to be linear and passing through coordinates origin. This means that the distance between nanofiller particle aggregates is limited by

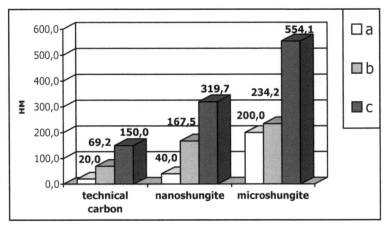

FIGURE 6.3   The initial particles diameter (a) their aggregates size in nanocomposite, (b) and distance between nanoparticles aggregates, (c) for nanocomposites, filled with technical carbon, nano, and microshungite.

Mean displacement of statistical walks, by which nanoparticles are simulated. The relationship between L_n and ξ can be expressed analytically as follows:

$$L_n \approx 9.6\xi \text{ nm} \tag{6.18}$$

The second important aspect of the model [31] in reference to nanofiller particle aggregation simulation is a finite nonzero initial particle concentration c or φ_n effect, which takes place in any real systems. This effect is realized at the condition $\xi \approx R_{ag}$, which occurs at the critical value R_{ag} (R_c), determined according to the relationship [31]:

$$c \sim R_c^{d_f^{ag}-d} . \tag{6.19}$$

The right-hand side of Eq. (6.19) represents cluster (particles aggregate) mean density. This equation establishes that fractal growth continues only, until cluster density reduces up to medium density, in which it grows. The calculated values R_c, according to Eq. (6.19), for the considered nanoparticles are adduced in Table 6.1, from which it follows that they give reasonable correspondence with this parameter experimental values R_{ag} (the average discrepancy of R_c and R_{ag} makes up 24%).

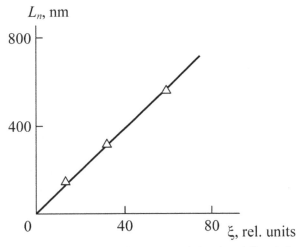

FIGURE 6.4 The relation between diffusion correlation length ξ and distance between nanoparticles aggregates L_n for considered nanocomposites.

As the treatment [31] was obtained within the frameworks of a more general model of diffusion-limited aggregation, its correspondence to the experimental data indicated unequivocally, such that aggregation processes in these systems were controlled by diffusion. Therefore, let us consider briefly the nanofiller particle diffusion. Statistical walkers diffusion constant ζ can be determined with the aid of the equation as in what follows [31]:

$$\xi \approx (\zeta t)^{1/2},\tag{6.20}$$

where t is walk duration.

Equation (6.20) supposes (at t = const) ζ increases in a number technical carbon—nanoshungite—microshungite as 196–1069–3434 relative units, i.e., diffusion intensification at diffusible particles size growth. At the same time, diffusivity D for these particles can be described by the well-known Einstein's relationship as in what follows [33]:

$$D = \frac{kT}{6\pi\eta r_n \alpha},$$ (6.21)

where k is Boltzmann constant, T is temperature, η is medium viscosity, α is numerical coefficient, which further is accepted equal to 1.

In its turn, the value η can be estimated according to equation as in what follows [34]:

$$\frac{\eta}{\eta_0} = 1 + \frac{2.5\phi_n}{1-\phi_n},$$ (6.22)

where η_0 and η are initial polymer and its mixture with nanofiller viscosity, respectively.

The calculation according to Eqs. (6.21) and (6.22) shows that within the aforementioned nanofiller number the value D changes as 1.32–1.14–0.44 relative units, i.e., reduces in three times, which was expected. This apparent contradiction is due to the choice of the condition $t = $ const (where t is nanocomposite production duration) in Eq. (6.20). In real conditions the value t is restricted by nanoparticle contact with growing aggregate and then instead of t the value t/c_0 should be used, where c_0 is the seeds concentration, determined according to Eq. (6.14). In this case, the value ζ for the indicated nanofillers changes as 0.288–0.118–0.086, i.e., it reduces in 3.3 times, which corresponds fully to the calculation according to the Einstein's relationship (Eq. (6.21)). This means that nanoparticle diffusion in polymer matrix obeys classical laws of Newtonian rheology [33].

Thus, the disperse nanofiller particle aggregation in elastomeric matrix can be described theoretically within the frameworks of a modified model of irreversible aggregation particle-cluster. The obligatory consideration of nanofiller initial particle size is a feature of the indicated model application to real system description. The indicated particles' diffusion in polymer matrix obeys classical laws of Newtonian liquids hydrodynamics. The offered approach allows to predict nanoparticle aggregate final parameters as a function of the initial particles' size, their contents, and other factors.

At present there are several methods of filler structure (distribution) determination in polymer matrix, both experimental [10, 35] and theoretical [4]. All the indicated methods describe this distribution by fractal dimension D_n of filler particle network. However, correct determination of any object fractal (Hausdorff) dimension includes three obligatory conditions. The first is the aforementioned determination of fractal dimension numerical magnitude, which should not be equal to object topological dimension. It is well known [36], that any real (physical) fractal possesses fractal properties within a certain scales range. Therefore, the second condition is the evidence of object self-similarity in this scale range [37]. Finally, the third condition is the correct choice of measurement scale range itself. As it has been shown in Refs. [38, 39], the minimum range should exceed at any rate one self-similarity iteration.

The first method of dimension D_n experimental determination uses the following fractal relationship [40, 41]:

$$D_n = \frac{\ln N}{\ln \rho},$$ (6.23)

where N is a number of particles with size ρ.

Particle sizes were established on the basis of atomic-power microscopy data (see Figure 6.2). For each from the three-studied nanocomposites no less than 200 particles were measured, the sizes of which were united into 10 groups and mean values N and ρ were obtained. The dependences $N(\rho)$ in double logarithmic coordinates were plotted, which proved to be linear and the values D_n were calculated according to their slope (see Figure 6.5).

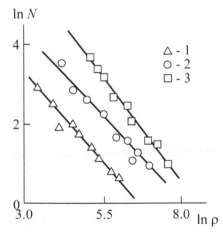

FIGURE 6.5   The dependences of nanofiller particles number N on their size ρ for nanocomposites BSR/TC (1) BSR/nanoshungite and (2) BSR/microshungite.

It is obvious that at such approach fractal dimension D_n is determined in two-dimensional Euclidean space, whereas real nanocomposite should be considered in three-dimensional Euclidean space. The following relationship can be used for D_n re-calculation for the case of three-dimensional space [42]:

$$D_3 = \frac{d + D_2 \pm \left[(d - D_2)^2 - 2 \right]^{1/2}}{2}, \tag{6.24}$$

where D_3 and D_2 are corresponding fractal dimensions in three- and two-dimensional Euclidean spaces, respectively, and $d = 3$.

The calculated dimensions D_n, based on the aforementioned method, are adduced in Table 6.2. The values D_n for the studied nanocomposites are varied within the range of 1.10–1.36, i.e., they characterize more or less branched linear formations ("chains") of nanofiller particles (aggregates of particles) in elastomeric nanocomposite structure. Let us remind that for particulate-filled composites polyhydroxyether/graphite, the value D_n changes within the range of ~2.30–2.80 [4, 10], i.e., for these materials filler particles network is a bulk object, but not a linear one [36].

TABLE 6.2 The dimensions of nanofiller particles (aggregates of particles) structure in elastomeric nanocomposites

Nanocomposite	D_n, Eq. (6.23)	D_n, the Eq. (6.25)	d_0	d_{surf}	φ_n	D_n, Eq. (6.29)
BSR/TC	1.19	1.17	2.86	2.64	0.48	1.11
BSR/nanoshungite	1.10	1.10	2.81	2.56	0.36	0.78
BSR/microshungite	1.36	1.39	2.41	2.39	0.32	1.47

Another method of D_n experimental determination uses the so-called "quadrates method" [43], which consists of the following parameters. On the enlarged nanocomposite microphotograph (see Figure 6.2), a net of quadrates with quadrate side size α_i, changing from 4.5 to 24 mm with constant ratio $\alpha_{i+1}/\alpha_i = 1.5$, is applied and then quadrates number N_i, in to which nanofiller particles hit (fully or partly), is counted up. Five arbitrary net positions concerning microphotograph were chosen for each measurement. If nanofiller particles' network is a fractal, then the following relationship should be fulfilled [43]:

$$N_i \sim S_i^{-D_n/2}, \tag{6.25}$$

where S_i is quadrate area, which is equal to α_i^2.

In Figure 6.6 the dependences of N_i on S_i in double logarithmic coordinates for the three-studied nanocomposites, corresponding to the relationship Eq. (6.25), is adduced. As one can see, these dependences are linear, which allows to determine the value D_n from their slope. The determined values D_n, according to Eq. (6.25), are also adduced in Table 6.2, from which a good correspondence of dimensions D_n, obtained by the two aforementioned methods, follows (their average discrepancy makes up to 2.1 per cent after these dimensions re-calculation for three-dimensional space according to Eq. (6.24)).

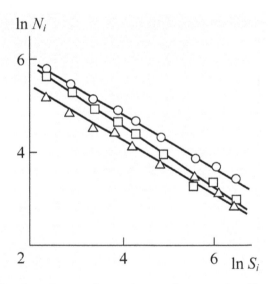

FIGURE 6.6   The dependences of covering quadrates number N_i on their area S_i, corresponding to the relationship Eq. (6.25), in double logarithmic coordinates for nanocomposites on the basis of BSR. The designations are the same, that in Figure 6.5.

As it has been shown in Ref. [44], the usage for self-similar fractal objects at Eq. (6.25) is based on the equation as in what follows:

$$N_i - N_{i-1} \sim S_i^{-D_n} \tag{6.26}$$

In Figure 6.7, the dependence, corresponding to Eq. (6.26), for the three-studied elastomeric nanocomposites is adduced. As one can see, this dependence is linear, passes through origin of coordinates, which according to Eq. (6.26) is confirmed by nanofiller particle (aggregates of particles) "chains" self-similarity within the selected α_i range. It is obvious that this self-similarity will be a statistical one [44]. Let us note that the points, corresponding to α_i=16 mm for nanocomposites butadiene—styrene rubber/technical carbon (BSR/TC) and butadiene—styrene rubber/microshungite (BSR/microshungite), do not correspond to a common straight line. Accounting for electron microphotographs of Figure 6.2 enlargement gives the self-similarity range for nanofiller "chains" of 464–1472 nm. For nanocomposite butadiene—styrene rubber/nanoshungite (BSR/nanoshungite), which has no points deviating from a straight line of Figure 6.7, α_i range makes up 311–1,510 nm, which corresponds well enough to the aforementioned self-similarity range.

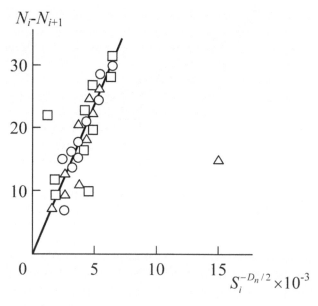

FIGURE 6.7 The dependences of $(N_i - N_i + 1)$ on the value $S_i^{-D_n/2}$, corresponding to the relationship Eq. (6.26), for nanocomposites on the basis of BSR. The designations are the same, that in Figure 6.5.

Measurement scales S_i minimum range should contain at least one self-similarity iteration as shown in Refs. . In this case the condition for ratio of maximum, S_{max}, and minimum, S_{min}, areas of covering quadrates should be fulfilled [39]:

$$\frac{S_{max}}{S_{min}} > 2^{2/D_n} . \qquad (6.27)$$

Hence, accounting for the aforementioned restriction let us obtain $S_{max}/S_{min} = 121/20.25 = 5.975$ that is larger than values $2^{2/D_n}$ for the studied nanocomposites, which are equal to 2.71–3.52. This means that measurement scales range is chosen correctly.

The self-similarity iterations number μ can be estimated from the inequality [39]:

$$\left(\frac{S_{max}}{S_{min}}\right)^{D_n/2} > 2^{\mu} . \qquad (6.28)$$

Using the aforementioned values of Eq. (6.28) parameters, $\mu = 1.42 - 1.75$ is obtained for the studied nanocomposites, i.e., in our experiment conditions self-similarity iterationnumber is larger than unity, which again confirms correctness of the value D_n estimation [35].

In addition, let us consider the physical grounds of smaller values D_n for elastomeric nanocomposites in comparison with polymer microcomposites, i.e., the causes of nanofiller particle (aggregates of particles) "chains" formation in the first ones. The value D_n can be determined theoretically according to the equation as in what follows [4]:

$$\phi_{if} = \frac{D_n + 2.55d_0 - 7.10}{4.18} , \qquad (6.29)$$

where φ_{if} is the relative fraction of interfacial regions, and d_0 is nanofiller initial particle surface dimension.

The dimension d_0 estimation can be carried out with the help of Eq. (6.4) and the value φ_{if} can be calculated according to Eq. (6.7). The results of dimension D_n theoretical calculation according to Eq. (6.29) are adduced in Table 6.2, from which a theory and experiment good correspondence follows. Eq. (6.29) indicates unequivocally to the cause of a filler in nano and microcomposites different behavior. The high (close to 3, see

Table 6.2) values d_0 for nanoparticles and relatively small ($d_0 = 2.17$ for graphite [4]) values d_0 for microparticles at comparable values φ_{if} are such cause for composites of the indicated classes [3, 4].

Hence, the aforementioned results have shown that nanofiller particle (aggregates of particles) "chains" in elastomeric nanocomposites are physical fractal within self-similarity (and, hence, fractality [41]) range of ~500–1,450 nm. In this range, their dimension D_n can be estimated according to Eqs. (6.23), (6.25), and (6.29). The cited examples demonstrate the necessity of the measurement scales range with correct choice.

As it has been noted earlier [45], the linearity of the plots, corresponding to Eqs. (6.23) and (6.25), and D_n nonintegral value do not guarantee object self-similarity (and, hence, fractality). The nanofiller particle (aggregates of particles) structure low dimensions are due to the initial nanofiller particles surface high fractal dimension.

In Figure 6.8, the histogram is adduced, which shows elasticity modulus E change, obtained in nanoindentation tests, as a function of load on indenter P or nanoindentation depth h. If the dependences $E(P)$ or $E(h)$ are identical qualitatively for all the three considered nanocomposites, then the dependence $E(h)$ for nanocomposite BSR/TC was chosen, which reflects the indicated scale effect quantitative aspect in the most clearest way.

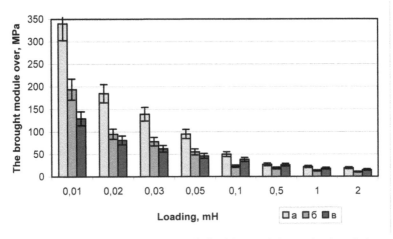

FIGURE 6.8   The dependences of reduced elasticity modulus on load on indentor for nanocomposites on the basis of butadiene—styrene rubber, (a) filled with technical carbon, (b) micro, and (c) nanoshungite.

In Figure 6.9, the dependence of E on h_{pl} (see Figure 6.10) is adduced, which breaks down into two linear parts. Such dependences on elasticity modulus—strain are typical for polymer materials in general and are due to intermolecular bonds anharmonicity [46].

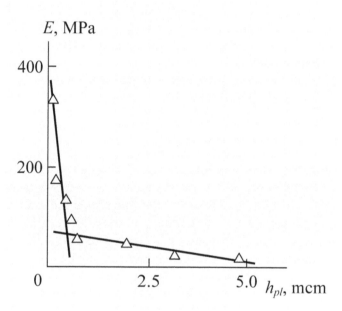

FIGURE 6.9   The dependence of reduced elasticity modulus E, obtained in nanoindentation experiment, on plastic strain h_{pl} for nanocomposites BSR/TC.

it has been shown in Ref. that the dependence $E(h_{pl})$ first part at $h_{pl} \leq$ 500 nm is not connected with relaxation processes and has a purely elastic origin. The elasticity modulus E on this part changes in proportion to h_{pl} as:

$$E = E_0 + B_0 h_{pl}, \qquad (6.30)$$

where E_0 is "initial" modulus, i.e., modulus, extrapolated to $h_{pl} = 0$, and the coefficient B_0 is a combination of the first and second kind elastic constants. In the considered case, $B_0 < 0$. Furthermore, Grüneisen parameter γ_L, characterizing intermolecular bonds anharmonicity level, can be determined as in what follows [47]:

$$\gamma_L \approx -\frac{1}{6} - \frac{1}{2}\frac{B_0}{E_0}\frac{1}{(1-2\nu)},$$ (6.31)

where ν is Poisson ratio, accepted for elastomeric materials equal to ~0.475 [36].

According to Eq. (6.31), the following values has been calculated: γ_L: 13.6 for the first part and 1.50—for the second. Let us note the first from γ_L adduced values is typical for intermolecular bonds, whereas the second value γ_L is much closer to the corresponding value of Grüneisen parameter G for intrachain modes [46].

Poisson's ratio ν can be estimated by γ_L (or G) known values according to the equation as in what follows [46]:

$$\gamma_L = 0.7\left(\frac{1+\nu}{1-2\nu}\right).$$ (6.32)

The estimated values according to Eq. (6.32) are $\nu = 0.462$ for the dependence $E(h_{pl})$ first part and $\nu = 0.216$ for the second one. If for the first part the value ν is close to Poisson's ratio magnitude for nonfilled rubber [36], then in the second part, the additional estimation is required. As it is known [48], a polymer composites (nanocomposites) Poisson's ratio value ν_n can be estimated according to the equation as in what follows:

$$\frac{1}{\nu_n} = \frac{\phi_n}{\nu_{TC}} + \frac{1-\phi_n}{\nu_m},$$ (6.33)

where φ_n is nanofiller volume fraction, ν_{TC} and ν_m are nanofiller (technical carbon) and polymer matrix Poisson's ratio, respectively.

The value ν_m is equal to 0.475 [36] and the magnitude ν_{TC} is estimated as follows [49]. As it is known [50], the nanoparticles TC aggregates fractal dimension d_f^{ag} value which is equal to 2.40 and then the value ν_{TC} can be determined according to equation [50]:

$$d_f^{ag} = (d-1)(1+\nu_{TC}).$$ (6.34)

where $\nu_{TC} = 0.20$ and ν_n according to Eq. (6.33) gives the value 0.283, which is close enough to the value $\nu = 0.216$ according to Eq. (6.32) estimation. The obtained values ν and ν_n comparison demonstrates that in the dependence $E(h_{pl})$ ($h_{pl} < 0.5$ mcm), the first part in nanoindentation tests rubber-like polymer matrix ($\nu = \nu_m \approx 0.475$) is included and the second part—the entire nanocomposite as homogeneous system $v = v_n \approx 0.22$ is included [51].

Let us consider reduction E at h_{pl} growth (Figure 6.9) within the frameworks of density fluctuation theory, which value ψ can be estimated as follows [22]:

$$\psi = \frac{\rho_n kT}{K_T},$$ (6.35)

where ρ_n is nanocomposite density, k is Boltzmann constant, T is testing temperature, and K_T is isothermal modulus of dilatation, connected with Young's modulus E by the relationship [46]:

$$K_T = \frac{E}{3(1-v)}.$$ (6.36)

The scheme of volume of the deformed at nanoindentation material V_{def} calculation in case of using Berkovich indentor is adduced in Figure 6.10 and the dependence $\psi(V_{def})$ in logarithmic coordinates was shown in Figure 6.11. As it follows from the data of Figure 6.11, the density fluctuation growth is observed at the increased deformed material volume. The plot $\psi(\ln V_{def})$ extrapolation to $\psi = 0$ gives $\ln V_{def} \approx 13$ or $V_{def}(V_{def}^{cr}) = 4.42 \approx 10^5$ nm³. Having determined the linear scale l_{cr} of transition to $\psi = 0$ as $(V_{def}^{cr})^{1/3}$, let us obtain $l_{cr} = 75.9$ nm, which is close to nanosystems dimensional range upper boundary [6], which is equal to 100 nm. Thus, the aforementioned results stated that nanosystems are such systems in which density fluctuations are absent, and always taking place in microsystems.

As it follows from the data of Figure 6.9, the transition from nano to microsystems occurs within the range $h_{pl} = 408–726$ nm. Both the aforementioned values h_{pl} and $(V_{def})^{1/3} \approx 814–1,440$ nm can be chosen as the linear length scale l_n, corresponding to this transition. Based on the comparison of these values l_n with the distance between nanofiller particle aggregates L_n ($L_n = 219.2 − 788.3$ nm for the considered nanocomposites, see Figure 6.3) it follows that for transition from nano to microsystems, l_n should include at least two nanofiller particle aggregates and surround them with layers of polymer matrix, which is the lowest linear scale of nanocomposite simulation as a homogeneous system. It is easy to see that nanocomposite structure homogeneity condition is harder than the aforementioned system obtained from the criterion $\psi = 0$. Let us note that such method, namely, a nanofiller particle and surrounding it polymer matrix layers separation, is widespread at a relationship derivation in microcomposite models.

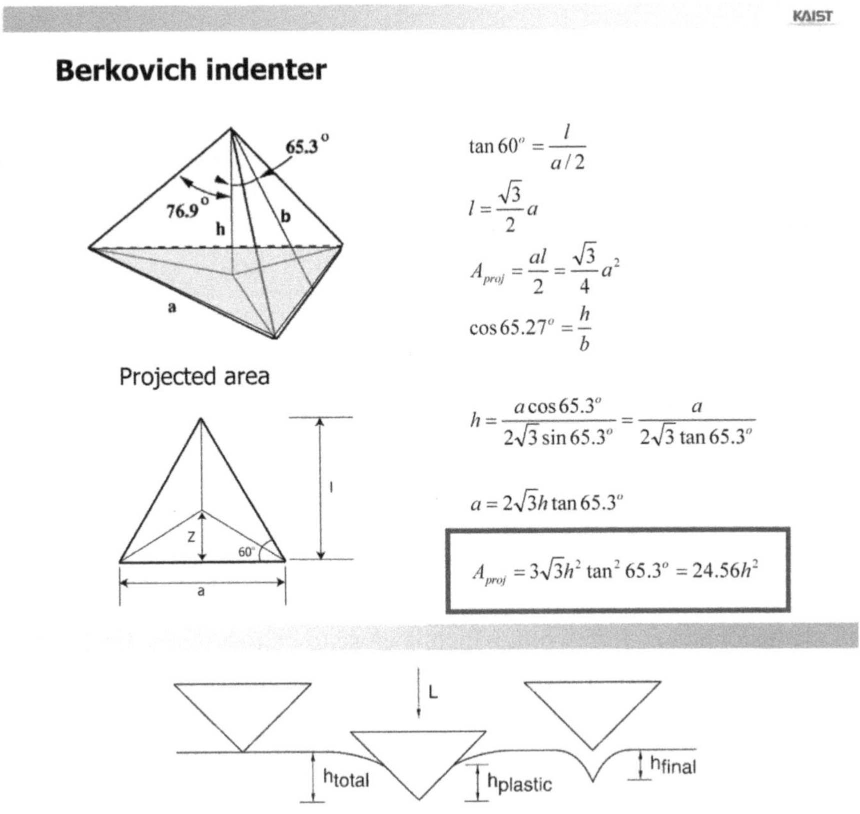
KAIST
Berkovich indenter
65.3°
76.9°
h
b
a
$\tan 60^{o} = \frac{l}{a/2}$
$l = \frac{\sqrt{3}}{2}a$
$A_{proj} = \frac{al}{2} = \frac{\sqrt{3}}{4}a^2$
$\cos 65.27^{o} = \frac{h}{b}$
Projected area
$h = \frac{a\cos 65.3^{o}}{2\sqrt{3}\sin 65.3^{o}} = \frac{a}{2\sqrt{3}\tan 65.3^{o}}$
$a = 2\sqrt{3}h\tan 65.3^{o}$
$A_{proj} = 3\sqrt{3}h^2\tan^2 65.3^{o} = 24.56h^2$
z
60°
l
L
htotal
hplastic
hfinal

FIGURE 6.10 The schematic image of berkovich indentor and nanoindentation process.

It is obvious that Eq. (6.35) is inapplicable to nanosystems, since $\psi \rightarrow 0$ assumes $K_T \rightarrow \infty$, which is physically incorrect. Therefore the value E_0, obtained by the dependence $E(h_{pl})$ extrapolation (see Figure 6.9) to $h_{pl} = 0$, should be accepted as E for nanosystems [49].

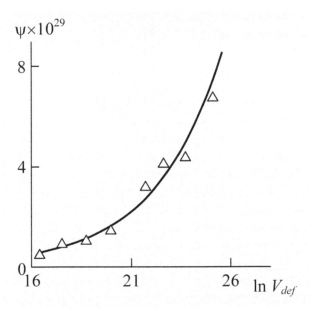

FIGURE 6.11   The dependence of density fluctuation ψ on volume of deformed in nanoindentation process material V_{def} in logarithmic coordinates for nanocomposites BSR/TC.

Hence, the aforementioned results have shown that elasticity modulus change at nanoindentation for particulate-filled elastomeric nanocomposites is due to a number of causes, which can be elucidated within the frameworks of anharmonicity conception and density fluctuation theory. Application of the first from the indicated conceptions assumes that in nanocomposites during nanoindentation process local strain is realized, affecting polymer matrix only, and the transition to macrosystems means nanocomposite deformation as homogeneous system.

The second from the mentioned conceptions has shown that nano- and microsystems differ by density fluctuation which does not exists in the first but in the second. The last circumstance assumes that for the considered nanocomposites, density fluctuations take into account nanofiller and polymer matrix density difference. The transition from nano- to microsystems is realized in the case when the deformed material volume exceeds nanofiller particle aggregate and surrounds it layers of polymer matrix combined volume [49].

The elastomeric nanocomposites reinforcement degree E_n/E_m description was derived as in what follows [3]:

$$\frac{E_n}{E_m} = 15.2\left[1-\left(d-d_{surf}\right)^{1/t}\right], \qquad (6.37)$$

where t is index percolation, equal to 1.7 [28].

From Eq. (6.37) it follows that nanofiller particle (aggregates of particles) surface dimension d_{surf} is the parameter, controlling nanocomposites reinforcement degree [53]. This postulate corresponds to the known principle about the decisive role of numerous division surfaces in nanomaterials as the basis of their properties change [54]. From Eqs. (6.4) to (6.6) it follows unequivocally that the value d_{surf} is defined by nanofiller particles (aggregates of particles) size R_p only. In its turn, from Eq. (6.37) it follows that elastomeric nanocomposites reinforcement degree E_n/E_m is defined by the dimension d_{surf}, the size R_p. This means that the reinforcement effect is controlled by nanofiller particles' (aggregates of particles) sizes and in virtue of this is the true nanoeffect.

In Figure 6.12, the dependence of E_n/E_m on $(d-d_{surf})^{1/1.7}$ is adduced, corresponding to Eq. (6.37), for nanocomposites with different elastomeric matrices (natural and butadiene—styrene rubbers, NR, and BSR) and different nanofillers (technical carbon of different marks, nano- and microshungite). Despite the indicated distinctions in composition, all adduced data are described well by Eq. (6.37).

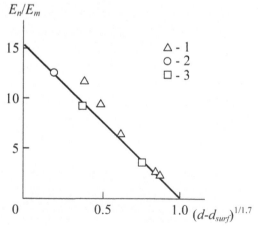

FIGURE 6.12   The dependence of reinforcement degree E_n/E_m on parameter $(d-d_{surf})^{1/1.7}$ value for nanocomposites (1) NR/TC, (2) BSR/TC, and (3) BSR/Shungite.

In Figure 6.13, two theoretical dependences of E_n/E_m on nanofiller particle size (diameter D_p), calculated according to Eqs. (6.4), (6.6), and (6.37), are adduced. However, at the curve 1, the calculated value D_p for the initial nanofiller particles was used, and at the curve 2 the value D_p^{ag} for—nanofiller particle aggregates size was used (see Figure 6.3). As it was expected [5], the growth E_n/E_m at D_p or D_p^{ag} reduction, along with the calculation of D_p (nonaggregated nanofiller), gives higher E_n/E_m values in comparison with the aggregated one (D_p^{ag} using). At $D_p \le 50$ nm faster growth E_n/E_m at D_p reduction is observed than at $D_p > 50$ nm, which was also expected. In Figure 6.13, the critical theoretical value D_p^{cr} for this transition, calculated according to the aforementioned general principles [52--54], is pointed out by a vertical shaded line. In conformity with these principles the nanoparticles size in nanocomposite is determined according to the condition, when division surface fraction in the entire nanomaterial volume makes up about ≥ 50 per cent.

This fraction is estimated approximately by the ratio $3l_{if}/D_p$, where l_{if} is interfacial layer thickness. As mentioned earlier, the data of Figure 6.1 gave the average experimental value $l_{if} \approx 8.7$ nm. Furthermore, from the condition $3l_{if}/D_p \approx 0.5$ let us obtain $D_p \approx 52$ nm, which is shown in Figure 6.13 by a vertical shaded line. As it was expected, the value $D_p \approx 52$ nm acts as a boundary between regions of slow ($D_p > 52$ nm) and fast ($D_p \le 52$ nm) E_n/E_m growth at D_p reduction. In other words, the materials with nanofiller particle size $D_p \le 52$ nm ("super-reinforcing" filler according to the terminology of Ref. [5]) should be considered as true nanocomposites.

Let us note in conclusion that although the curves 1 and 2 of Figure 6.13 are similar, nanofiller particle aggregation, which the curve 2 accounts for, reduces essentially enough nanocomposite reinforcement degree. At the same time. the experimental data correspond exactly to the curve 2 that was to be expected in virtue of aggregation processes, which always took place in real composites [4] (nanocomposites [52--55]). The values d_{surf}, obtained according to Eqs. (6.4)–(6.6), correspond well to the experimentally determined ones. Therefore, for nanoshungite and two marks of technical carbon, the calculation by the aforementioned method gives the following d_{surf} values: 2.81, 2.78, and 2.73, whereas experimental values of this parameter are equal to: 2.81, 2.77, and 2.73; i.e., practically a full correspondence of theory and experiment was obtained.

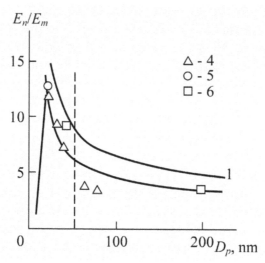

FIGURE 6.13   The theoretical dependences of reinforcement degree E_n/E_m on nanofiller particles size D_p, calculated according to Eqs. (6.4), (6.6), and (6.37), at initial nanoparticles (1) and nanoparticles aggregate (2) size using. The boundary value D_p (3), corresponding to true nanocomposite. Experimental data for nanocomposites NR/TC (4), BSR/TC (5), and BSR/shungite (6).

6.4   CONCLUSIONS

Hence, the stated aforementioned results have shown that the elastomeric reinforcement effect is the true nanoeffect, which is defined by the initial nanofiller particle size only. The indicated particle aggregation, always taking place in real materials, changes and reduces reinforcement degree quantitatively. This effect theoretical treatment can be received within the frameworks of fractal analysis. For the considered nanocomposites the nanoparticle size upper limiting value makes up ~ 52 nm.

KEYWORDS

- **Nanocomposites**
- **Nanofillers**
- **Polymer reinforcement**

REFERENCES

1. Yanovskii, Yu. G.; Kozlov, G. V.; and Karnet, Yu. N.; Mekhanika Kompozitsionnykh Materialov i Konstruktsii, **2011**, *17(2)*, 203–208.
2. Malamatov, A. Kh.; Kozlov, G. V.; and Mikitaev, M. A.; Reinforcement Mechanisms of Polymer Nanocomposites. Moscow: Publishers of the D.I. Mendeleev RKhTU; **2006**, 240 p.
3. Mikitaev, A. K.; Kozlov, G. V.; and Zaikov, G. E.; Polymer Nanocomposites: Variety of Structural Forms and Applications. Moscow: Nauka; **2009**, 278 p.
4. Kozlov, G. V.; Yanovskii, Yu. G.; and Karnet, Yu. N.; Structure and Properties of Particulate-Filled Polymer Composites: The Fractal Analysis. Moscow: Al'yanstransatom; **2008**, 363 p.
5. Edwards, D. C. J.; *Mater. Sci.* **1990**, *25(12)*, 4175–4185.
6. Buchachenko, A. L.; *Uspekhi Khimii.* **2003**, *72(5)*, 419–437.
7. Kozlov, G. V.; Yanovskii, Yu. G.; Burya, A. I.; and Aphashagova, Z. Kh.; *Mekhanika Kompozitsionnykh Materialov i Konstruktsii.* **2007**, *13(4)*, 479–492.
8. Lipatov, Yu. S.; The Physical Chemistry of Filled Polymers. Moscow: Khimiya; **1977**, 304 p.
9. Bartenev, G. M.; and Zelenev, Yu. V.; Physics and Mechanics of Polymers. Moscow: Vysshaya Shkola; **1983**, 391 p.
10. Kozlov, G. V.; and Mikitaev, A. K.; *Mekhanika Kompozitsionnykh Materialov i Konstruktsii.* **1996**, *2(3–4)*, 144–157.
11. Kozlov, G. V.; Yanovskii, Yu. G.; and Zaikov, G. E.; Structure and Properties of Particulate-Filled Polymer Composites: The Fractal Analysis. New York: Nova Science Publishers Inc.; **2010**, 282 p.
12. Mikitaev, A. K.; Kozlov, G. V.; and Zaikov, G. E.; Polymer Nanocomposites: Variety of Structural Forms and Applications. New York: Nova Science Publishers Inc.; **2008**, 319 p.
13. McClintok, F. A.; and Argon, A. S.; Mechanical Behavior of Materials. Reading, Addison-Wesley Publishing Company Inc.; **1966**, 440 p.
14. Kozlov, G. V.; Mikitaev, A. K.; and Doklady, A. N.; *SSSR.* **1987**, *294(5)*, 1129–1131.
15. Honeycombe, R. W. K.; The Plastic Deformation of Metals. Boston: Edward Arnold Publishers Ltd.; **1968**, 398 p.
16. Dickie, R. A.; In: Book: Polymer Blends. New York, San-Francisco, London: Academic Press; **1980**, *1*, 386–431 p.
17. Kornev, Yu. V.; Yumashev, O. B.; Zhogin, V. A.; Karnet, Yu. N.; and Yanovskii, Yu. G.; *Kautschuk i Rezina.* **2008**, *6*, 18–23.
18. Oliver, W. C.; and Pharr, G. M. J.; *Mater. Res.* **1992**, *7(6)*, 1564–1583.
19. Kozlov, G. V.; Yanovskii, Yu. G.; and Lipatov, Yu. S.; *Mekhanika Kompozitsionnykh Materialov i Konstruktsii.* **2002**, *8(1)*, 111–149.
20. Kozlov, G. V.; Burya, A. I.; and Lipatov, Yu. S.; *Mekhanika Kompozitnykh Materialov.* **2006**, *42(6)*, 797–802.
21. Hentschel, H. G. E.; and Deutch, J. M.; *Phys. Rev. A.* **1984**, *29(3)*, 1609–1611.
22. Kozlov, G. V.; Ovcharenko, E. N.; and Mikitaev, A. K.; Structure of Polymers Amorphous State. Moscow: Publishers of the D.I. Mendeleev RKhTU; **2009**, 392 p.

23. Yanovskii, Yu. G.; and Kozlov, G. V.; *Mater. VII Int. Sci. Pract. Conf. "New Polym. Compos. Mater." Nal'chik, KBSU.* **2011**, 189–194.
24. Wu, S. J.; *Polym. Sci.: Part B: Polym. Phys.* **1989**, *27(4)*, 723–741.
25. Aharoni, S. M.; Macromolecules. **1983**, *16(9)*, 1722–1728.
26. Budtov, V. P.; The Physical Chemistry of Polymer Solutions. Khimiya: Sankt-Peterburg; **1992**, 384 p.
27. Aharoni, S. M.; Macromolecules. **1985**, *18(12)*, 2624–2630.
28. Bobryshev, A. N.; Kozomazov, V. N.; Babin, L. O.; and Solomatov, V. I;. Synergetics of Composite Materials. Lipetsk: NPO ORIUS; **1994**, 154 p.
29. Kozlov, G. V.; Yanovskii, Yu. G.; and Karnet, Yu. N.; *Mekhanika Kompozitsionnykh Materialov i Konstruktsii.* **2005**, *11(3)*, 446–456.
30. Sheng, N.; Boyce, M. C.; Parks, D. M.; Rutledge, G. C.; Abes, J. I.; and Cohen, R. E.; *Polym.* **2004**, *45(2)*, 487–506.
31. Witten, T. A.; and Meakin, P.; *Phys. Rev. B.* **1983a**, *28(10)*, 5632–5642.
32. Witten, T. A.; and Sander, L. M.; *Phys. Rev. B.* **1983b**, *27(9)*, 5686–5697.
33. Happel, J.; and Brenner, G.; Hydrodynamics at Small Reynolds Numbers. Moscow: Mir; **1976**, 418 p.
34. Mills, N. J. J.; *Appl. Polym. Sci.* **1971**, *15(11)*, 2791–2805.
35. Kozlov, G. V.; Yanovskii, Yu. G.; and Mikitaev, A. K.; *Mekhanika Kompozitnykh Materialov.* **1998**, *34(4)*, 539–544.
36. Balankin, A. S.; Synergetics of Deformable Body. Moscow: Publishers of Ministry Defence SSSR; **1991**, 404 p.
37. Hornbogen, E.; *Int. Mater. Res.* **1989**, *34(6)*, 277–296.
38. Pfeifer, P.; *Appl. Surf. Sci.* **1984**, *18(1)*, 146–164.
39. Avnir, D.; Farin, D.; and Pfeifer, P. J.; *Colloid Interface Sci.* **1985**, *103(1)*, 112–123.
40. Ishikawa, K. J.; *Mater. Sci. Lett.* **1990**, *9(4)*, 400–402.
41. Ivanova, V. S.; Balankin, A. S.; Bunin, I. Zh.; and Oksogoev, A. A.; Synergetics and Fractals in Material Science. Moscow: Nauka; **1994**, 383 p.
42. Vstovskii, G. V.; Kolmakov, L. G.; and Terent'ev, V. E.; *Metally.* **1993**, *4*, 164–178.
43. Hansen, J. P.; and Skjeitorp, A. T.; *Phys. Rev. B.* **1988**, *38(4)*, 2635–2638.
44. Pfeifer, P.; Avnir, D.; and Farin, D. J.; *Stat. Phys.* **1984**, *36(5/6)*, 699–716.
45. Farin, D.; Peleg, S.; Yavin, D.; and Avnir, D.; *Langmuir.* **1985**, *1(4)*, 399–407.
46. Kozlov, G. V.; and Sanditov, D. S.; Anharmonical Effects and Physical-Mechanical Properties of Polymers. Novosibirsk: Nauka; **1994**, 261 p.
47. Bessonov, M. I.; and Rudakov, A. P.; *Vysokomolek. Soed. B.* **1971**, *13(7)*, 509–511.
48. Kubat, J.; Rigdahl, M.; and Welander, M. J.; *Appl. Polym. Sci.* **1990**, *39(5)*, 1527–1539.
49. Yanovskii, Yu. G.; Kozlov, G. V.; Kornev, Yu. V.; Boiko, O. V.; and Karnet, Yu. N.; *Mekhanika Kompozitsionnykh Materialov i Konstruktsii.* **2010**, *16(3)*, 445–453.
50. Yanovskii, Yu. G.; Kozlov, G. V.; and Aloev, V. Z.; *Mater. Int. Sci. -Pract. Conf. "Modern Problems of APK Innovation Dev. Theory Practice".* Nal'chik, KBSSKhA. **2011**, 434–437.
51. Chow, T. S.; *Polym.* **1991**, *32(1)*, 29–33.
52. Ahmed, S.; and Jones, F. R. J.; *Mater. Sci.* **1990**, *25(12)*, 4933–4942.
53. Kozlov, G. V.; Yanovskii, Yu. G.; and Aloev, V. Z.; *Mater. Int. Sci. -Pract. Conf. Dedicated to FMEP 50-th Anniversary.* Nal'chik, KBSSKhA. **2011**, p. 83–89.

54. Andrievskii, R. A.; *Rossiiskii Khimicheskii Zhurnal.* **2002**, *46(5),* 50–56.
Kozlov, G. V.; Sultonov, N. Zh.; Shoranova, L. O.; and Mikitaev, A. K.; *Naukoemkie Tekh-nologii.* **2011**, *12(3),* 17–22.

CHAPTER 7

A STUDY ON THE EFFECTS OF THE MODIFIED SILICA–GELATIN HYBRID SYSTEMS ON THE PROPERTIES OF PAPER PRODUCTS

PRZEMYSŁAW PIETRAS, ZENON FOLTYNOWICZ, HIERONIM MACIEJEWSKI, and RYSZARD FIEDOROW

CONTENTS

7.1   INTRODUCTION

Since the beginning of papermaking in Western Europe, gelatin was used to reduce the sorption of water (paper sizing) by papers to improve the buffer effect and feathering of the inks [1]. Nowadays, papers are sized with synthetic sizes such as alkyl ketene dimers (AKDs) and alkenyl succinic anhydrides (ASAs), which were developed for the paper industry in 1953 and 1974, respectively [2], but gelatin sizing continues to be used for artist quality papers.

Hybrid materials consisting of biopolymer and silica raise hopes for obtaining materials of high compatibility to tissues and interesting properties. Such hybrids find application as biocompatible materials, bone substituents, as well as immobilizers of enzymes, catalysts and sensors, pre-ceramic materials, and many others [3, 5, 6]. A very convenient and effective method for preparing such systems is sol–gel technique. Biocomposites obtained by the aforementioned method have exceptional properties such as high plasticity, low modulus of elasticity, and high strength [3–7].

Gelatin is a product of thermal denaturation of collagen presents in bones and animal skins. It is formed as a result of temperature action on three-chain polypeptide helix of collagen which undergoes unwinding to give gelatin balls. It is not a homogeneous product and its composition depends on collagen origin. Molecular weight of gelatin ranges between 80,000 and 2,00,000. Gelatin is a fully biodegradable material [8, 11]. It dissolves in water after heating to about 40°C. Gelatin has its greatest application in food, pharmaceutical, and photographic industries. Its potential applications include bone reconstructions, biocatalysis, drug distribution control in living organisms, and others [3, 12].

In the case of hybrid materials, the formation of durable bonds between biopolymer and inorganic gel is of great importance. The fundamental role in the formation of such bonds is played by silane coupling agents such as aminosilanes, glycidoxysilanes, and others [13]. For gelatin-containing systems the most frequently selected coupling agent is glycidoxypropyltrimethoxysilane. Ren et al. [14, 15] described synthesis of porous siloxane derivatives of gelatin and 3-glycidoxypropyltrimethoxysilane (GPTMS), prepared by sol–gel technique in the presence of HCl. Furthermore, Ren et al. [16] described synthesis of gelatin-siloxane hybrids with gelatin bound terminally to siloxane chains, using GPTMS as a coupling agent. Smitha et al. [17] presented a method for immobilization of gelatin molecules in mesoporous silica by carrying out tetraethoxysilane (TEOS) hydrolysis under conditions of controlled pH. Pietras et al. [18] described preparation of silica–gelatin hybrid materials by sol–gel method using ethyl silicate (U740) as a silica precursor and GPTMS as a coupling agent.

In this chapter we present syntheses of silica–gelatin hybrid materials aimed at obtaining possibly the highest degree of compatibility of the systems prepared by the integration of two types of polymers–organic biopolymer (gelatin) and inorganic polymer (silica gel) using GPTMS as a coupling agent. The obtained hybrid materials were applied as modifiers of paper products. The modified paper products were characterized from the point of view of the influence of employed impregnants on functional parameters such as basis weight change, durability on elongation, durability on rupture, absorptiveness of water, and burning time.

7.2 MATERIALS AND METHODS

7.2.1 MATERIALS

Gelatin employed in the study originated from porcine skin (type A, isoelectric point (pI)=8, 300 Bloom, dry matter 90%) was purchased from Aldrich (research grade gelatin (RGG)) and edible porcine gelatin (food grade gelation (FGG)) was purchased from Kandex, Ltd.. Ethyl silicate-U740 (40% SiO_2), TEOS, and GPTMS were obtained from Unisil Ltd., (Tarnów, Poland) and glacial acetic acid from Aldrich.

The modification was performed on white paper Lux 80 g/m² for black and white printing and copying, manufactured by International Paper , Poland, as well as cardboard for painting and building purposes, manufactured by Blue Dolphin Tapes, Poland. To compare properties of the impregnated paper materials with those commercially available, we have also subjected to impregnation of a packing paper manufactured by Hoomark, Ltd., Jędrzychowice, Poland (**PP**) and a cardboard A1 170 g/m² manufactured by Kreska, Bydgoszcz, Poland (**TZ**).

7.2.2   SILICA–GELATIN HYBRID MATERIALS

The starting sol was obtained by hydrolysis and condensation of an appropriate amount of ethyl silicate (U740, d = 1.05 g/cm^3, 40% SiO$_2$) or TEOS (d = 0.933 g/cm^3, 28.85% SiO$_2$) and GPTMS in water in the presence of acetic acid as pH control agent. In the typical experiment 166 g of TEOS or 120 g of U740, 61 g of GPTMS, and 60 g of acetic acid were used. The amount of silica formed as a result of hydrolysis and condensation of both precursors was constant and taken into account each time when proportions of ethyl silicate or TEOS to GPTMS were chosen. The mixture was vigorously stirred and heated at 85°C for 6 h to yield functionalized silica gel. At the next stage, gelatin was added to the obtained sol and the mixture was heated at 60°C for 2 h with vigorous stirring. Codes and compositions of samples obtained are shown in Table 7.1. Then this gel was applied as the impregnant for paper products.

TABLE 7.1   Percentage of gelatin* and codes of silica–gelatin hybrid materials based on TEOS and U740

Gelatin added [%]	TEOS-based sample coding	U740-based sample coding
0	T0	U0
50	T1	U1
100	T2	U2
150	T3	U3

*Calculated with reference to theoretical amount of silica formed from TEOS or U740.

7.2.3   MODIFICATION OF PAPER PRODUCTS

A 350 ml of the impregnant gel (ca. ¼ of obtained silica–gelatin hybrid system) was placed into a photographic tray. Six sheets of paper of A4 size or cardboard were modified in a given portion of the impregnant. The impregnation was carried out by deep coating method for 5 min (FGG5 and RGG5) or for 15 min (FGGS15 and RGG15). Then, the impregnated sample was hung over the tray to enable the excess impregnant to flow away. This was followed by drying of the sample in an air recirculation oven at 60°C for 30 min. The preparations obtained were designated as FGG5, FGG15, RGG5, and RGG15, respectively.

7.2.4 CHARACTERIZATION OF THE SYSTEMS

The obtained materials were characterized from the point of view of changes in functional parameters such as basis weight change, durability on elongation, durability on rupture, absorptiveness of water, and burning time.

Changes in grammage were measured as follows: samples of 1,000 mm^2 area were cut out of modified paper with the accuracy of 0.5 mm^2. Each sample was weighed on a balance with the accuracy of three significant digits. Measurements were performed for 10 samples of each modified paper and results were averaged. Durability on elongation was determined using tensile testing machine VEB TIW Rauenetein ZT; working length was 100 mm, sample width 15 mm, elongation speed 20 mm/ min. Durability on rupture of cardboards weighing over 250 g/cm^2 was measured with the use of an apparatus made by Lorentzen & Wettre Co. at initial pressure of 0.5 MPa. Absorptiveness of water was measured by the Cobb method in compliance with the standard ASTM D3285-93 (2005) by determining the Cobb60 index; sample surface area was 100 cm^2. Burning time was measured under controlled conditions in a closed chamber equipped with a burner for combustion of samples; height of the burner flame was 1.5–2 cm.

7.3 RESULTS AND DISCUSSION

7.3.1 CHARACTERIZATION OF MODIFIED PAPER PRODUCTS

Within the study two series were synthesized of silica–gelatin hybrid systems based on TEOS and U740 with increasing gelatin (FGG or RGG) content. The prepared silica–gelatin hybrid systems were used for the impregnation of paper products PB, TM, PP, and TZ to evaluate the effect of silica precursors applied in the study, as well as that of origin and concentration of gelatin on selected qualitative and functional parameters of the paper products studied.

All samples of paper products maintained their initial shape and size after impregnation and drying and no curling up occurred. However, an increase in the rigidity of samples was observed. The use of the impregnant had no influence whatsoever on the possibility of writing with a ball pen, pencil, or felt-tip pen on the samples. Organoleptic evaluation with naked eye did not show color changes in the investigated paper samples, except for TM in the case of which color darkening was observed. No separation of impregnant from paper was noticed during cutting of modified paper samples into smaller pieces required for different analyses. The samples have retained their shape after cutting they did not curl up.

7.3.2   BASIS WEIGHT CHANGE

Changes in the weightage of PB samples modified with hybrid systems based on TEOS as well as FGG and RGG were shown in Figure 7.1. One can notice that weightage of a given modified paper increases when compared with that of non-modified (NM) paper. Analogous situation was observed for all paper samples impregnated with all hybrid systems studied. Initial weightages of investigated papers were 80 g/m^2 for PB, 289 g/m^2 for TM, 51 g/m^2 for PP and 170 g/m^2 for TZ. The highest values of impregnated paper weightage were observed for the last samples of a given series modified with systems based on TEOS and U740 with the greatest FGG or RGG gelatin content. The weightage of PB increased to reach the highest value of 110 g/m^2 for the sample T3 FGG5. In the case of TM, analogously to the aforementioned paper samples, the weightage raises with increasing gelatin content to the maximum value of 363 g/m^2 in the case of the last sample of FGG15 series based on TEOS. The raise in weightage with increasing gelatin content in impregnants was also observed for all control paper samples. In the case of PP, the weightage increases to 71 g/m^2, whereas for TZ it increases to 194 g/m^2. The increase in weightage of all samples of all series of papers modified with hybrid materials when compared with starting papers testifies for the penetration of impregnants used in the papers investigated. This fact is additionally proved by the appearance and behavior of samples during their preparation and analyses (impregnant was not coming off paper during cutting out and no splinters were formed while testing strength of samples).

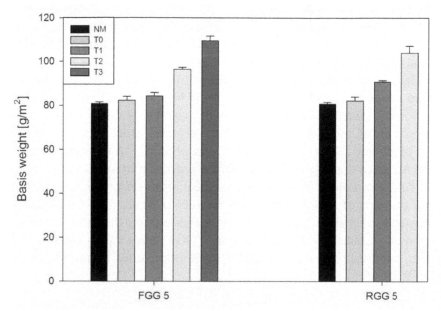

FIGURE 7.1   Weightage of non-modified (NM) PB and TEOS-modified (T0-T3) PB.

7.3.3   DURABILITY ON ELONGATION

Durability on elongation is the maximum tensile force acting on width unit of a sample at which the sample remains unbroken. The measurement of the aforementioned parameter consists of placing a sample in holders of tensile testing machine and stretching the sample with a constant speed until the latter underwent breaking. The value of tensile force at the moment of breaking was recorded. Averaged results of measurements of breaking resistance of PB impregnated with hybrid materials based on TEOS and FGG as well as RGG were shown in Figure 7.2. In the case of PB and TM, an improvement in this parameter was observed, whereas in that of PP and TZ a slight reduction in durability on elongation was noticed. Values of the aforementioned parameter were 3.8, 5.4, 6.2, and 9.4 kN/m for unmodified PB, TM, PP, and TZ, respectively.

The durability on elongation of impregnated PB increased with a rise in gelatin content for all series of both hybrid systems. The highest values were observed for the last samples of a given series and were 6.0, 6.2, 5.3, and 5.1 kN/m for FGG5 and RGG5 series for hybrid systems based on TEOS and U740, respectively. Similarly as it was in the case of PB, also in that of TM an increase was observed in the durability on elongation.

The aforementioned parameter has the highest value for T3 sample of FGG5 series and equals to 9.1 kN/m for hybrid systems based on TEOS and for U3 sample of FGG5 series being equal to 9.4 kN/m for hybrid systems based on U740. In contrast to previous paper products, in the case of PP and TZ, a small decrease was observed in the durability on elongation. In the case of PP, the highest reduction in this parameter occurs for T1 sample of RGG15 series and equals to 4.5 kN/m, whereas for other samples of all series it ranges between 5.2 and 6.0 kN/m. Results obtained for TZ show that the highest decrease in the durability on elongation takes place for T2 sample from FGG5 series and is equal to 7.8 kN/m, whereas for most of samples from other series it is in the range of 8–9 kN/m.

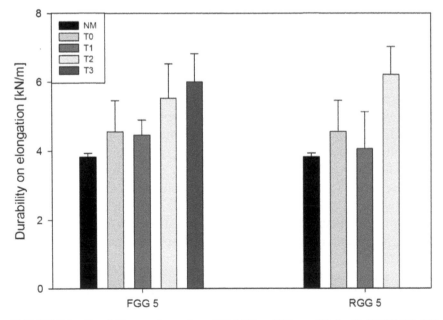

FIGURE 7.2 Durability on elongation of NM PB and PB modified with TEOS (T0–T3).

7.3.4   DURABILITY ON RUPTURE

Burst is the strength of a single sheet of paper to withstand homogeneous pressure acting perpendicularly to the surface of the sheet. It is the basic parameter for evaluation of strength of packing paper and corrugations of corrugate board. The measurement consists in tightening a paper sample between two rings of flexible membrane which undergoes bulging under the effect of pressure of air delivered until the sample breaks. Bursting strength of a sample corresponds to maximum value of pressure that resulted in the sample breakage. The measurement was carried out for 10 TM-modified samples and results were averaged. Figure 7.3 shows averaged results for TM modified with hybrid systems based on U740 and FGG as well as RGG. It is clearly seen in the figure that bursting strength of impregnated TM is considerably greater than that of non-impregnated TM for which the burst was 252 kPA. In the case of cardboard modified with hybrid systems based on U740 an increase in bursting strength is noticeable for both FGG series with the rise in gelatin content.

The values change from 350 to 531 kPa and 359 to 482 kPa for FGG5 and FGG15 series, respectively. For both RGG series, a decrease in bursting pressure was observed in the case of gelatin-containing samples when compared with samples impregnated with pure silica sol; however, the decrease was small (from 539.0 to 514.5 kPa and from 414.5 kPa do 393.0 kPa for RGG 5 and RGG15 series, respectively). In the case of samples modified with TEOS-based hybrid systems the bursting strength increased as much as to 555.5 kPa for samples from FGG5 and RGG5 series and to 548.0 kPa for those from RGG15 series. Finally, one can conclude that the impregnation of TM with the studied hybrid systems resulted in the increase of bursting strength which in some cases was over 100 per cent.

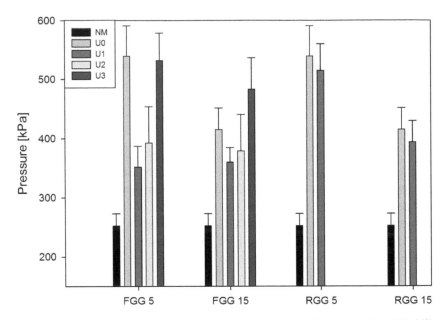

FIGURE 7.3   Durability on rupture of NM TM and TM modified with U740 (U0–U3).

7.3.5   ABSORPTIVENESS OF WATER

Absorptiveness of water is the mass of water absorbed by 1 m^2 of paper at a specified time and specified conditions. The measurement consists in determining the mass of water absorbed by a paper sample of 100 cm^2 area in a specified time under the water level of 1 cm. Measurements were performed for 10 samples of each modified paper and results were averaged.

The change in absorptiveness of water for PB samples modified with hybrid systems based on TEOS and FGG as well as RGG was presented in Figure 7.4. In the case of PB and TM an improvement in this parameter was observed, whereas in that of PP and TZ a considerable increase occurred in water absorptiveness. The relevant values were 95 g/m^2, 554 g/m^2, 30 g/m^2 and 58.5 g/m^2 for NM PB, TM, PP, and TZ, respectively. Reduction in the absorptiveness of water was observed for impregnated PB with increase in gelatin content for all series of both hybrid systems. The smallest value was found for T3 sample of TEOS-based RGG5 systems (64 g/m^2). In the case of TM, the water absorptiveness decreased for all samples modified with all hybrid systems studied with increased gelatin content, similarly as it was in the case of PB.

The lowest water absorptiveness (169 g/m^2) was found for the last sample of FGG5 series based on U740. In contrast to previously tested paper samples, all samples of modified PP and TZ were characterized by increased absorptiveness of water. In the case of PP, the greatest increase in this parameter occurred for all T0 samples of all series based on TEOS (the relevant values: 44, 47, 44, and 47 g/m^2) and for all U0 samples of all series of systems based on U740 (the relevant values: 37, 48, 37, and 48 g/m^2). For all TZ samples, the absorptiveness of water increased with the rise in gelatin content. The highest values of this parameter were found for last samples of all series for both hybrid systems. They were 195, 200, 182, and 182 g/m^2 for FGG5, FGG15, RGG5, and RGG15 series of TEOS-based systems and 167, 181, 190, and 198 g/m^2 for FGG5, FGG15, RGG5, and RGG15 series of U740-based systems.

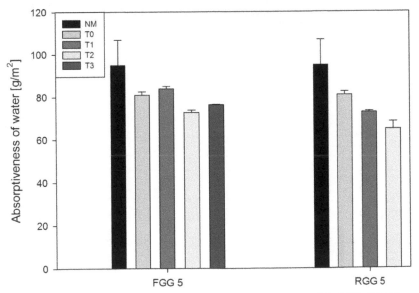

FIGURE 7.4 Absorptiveness of water of NM PB and TEOS-modified (T0-T3) PB.

7.3.6 BURNING TIME

The measurement of this parameter consists in placing a paper sheet sample in the flame of a burner and determining the burning time. (The burner was removed after the sample caught fire.) Ten samples of each modified paper were subjected to the measurement and results were averaged. The averaged results of burning time measurements for PB modified with hybrid systems based on U740 and FGG as well as RGG are given in Figure 7.5. Burning time for all series of modified paper samples was longer when compared with non-impregnated (NM) paper samples for which it was 9, 44, 9, and 21 s for PB, TM, PP, and TZ, respectively. In the case of PB the longest burning time (21 s) was observed for a sample from FGG5 series of TEOS-based system. In addition, in the case of TM an extension of burning time occurred with increasing gelatin content in all series of hybrid systems based on TEOS and U740. The longest burning time was 54 s for last samples from FG15 and RGG5 series of TEOS-based systems. An increase in burning time was also observed in the case of PP for which it was 17 s (for T3 sample from FGG5 series of TEOS-based systems) and in the case of TZ for which it was 35 s (for samples of TEOS-based systems).

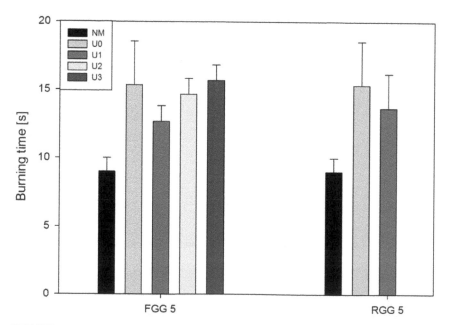

FIGURE 7.5   Burning time of NM PB and PB modified with U740 (U0-U3).

7.4 CONCLUSIONS

Results of the study conclude that the kind of silica precursor (TEOS or ethyl silicate) has no significant effect on upgrading selected functional properties of paper products. More important parameters appeared to be the kind and amount of gelatin added. Such parameters as weightage, absorptiveness of water as determined by the Cobb method, bursting strength, and burning time depend on gelatin content. A greater change in the studied parameters was observed in the case of RGG gelatin when compared with food grade gelatin. It was also established that the hybrid materials applied as impregnants for paper products either considerably (TM) or slightly (PB) improve functional parameters of materials NM by manufacturers during the production process. In the case of paper products (PP, TZ) pre-modified in their production process, no changes or a slight deterioration of the studied parameters were observed.

KEYWORDS

- **Ethyl silicate**
- **Gelatin**
- **Paper**
- **Paper product modification**
- **Sol–gel**
- **Tetraethoxysilane**

REFERENCES

1. Dupont, A. L.; *J. Chromatogr. A.* **2002,** *950,* 113–124.
2. Roberts, J. C.; Paper Chemistry. 2nd Ed. Blackie: Glasgow; **1996.**
3. Smitha, S.; Shajesh, P.; Mukundan, P.; Nair, T. D. R.; Warrier, K. G. K.; Colloids and Surfaces B: Biointerfaces 55. **2007,** 38–43.
4. Watzke, H. J.; Dieschbourg, C.; *Adv. Colloid Interface Sci.* **1994,** *50,* 1.
5. Schuleit, M.; Luisi, P. L.; *Biotechnol. Bioeng.* **2001,** *72,* 249.
6. Liu, D. M.; Chen, I. W.; *Acta Mater.* **1999,** *47, 4535.*
7. Yuan, G. L.; et al. *Chem.* **2000,** *159,* 45.
8. Boanini, E.; Rubini, K.; Panzavolta, S.; Bigi, A.; *Acta Biomater.* **2010,** *6,* 383–388.

9. Bigi, A.; Cojazzi, G.; Panzavolta, S.; Roveri, N.; and Rubini, K.; *Biomater.* **2002,** *23,* 4827–4832.

10. Karim, A. A.; and Bhat, R.; *Food Hydrocolloids.* **2009,** *23,* 563–576.

11. Bigi, A.; Cojazzi, G.; Panzavolta, S.; Rubini, K.; and Roveri, N.; *Biomater.* **2001,** *22,* 763–768.

12. Ren, L.; Tsuru, K.; Hayakawa, S.; Osaka, A.; *Biomater.* **2002,** *23,* 4765–4773.

13. Advincula, M.; Fan, X.; Lemons, J.; Advincula, R.; *Colloids Surf.: B,* **2005,** *42, 29.*

14. Ren, L.; Tsuru, K.; Hayakawa, S.; and Osaka, A.; *Biomater.* **2002,** *23, 4765.*

15. Ren, L.; Tsuru, K.; Hayakawa, S.; Osaka, A.; *J. Non-Cryst. Solids.* **2001a,** *285, 116.*

16. Ren, L.; Tsuru, K.; Hayakawa, S.; and Osaka, A.; *J. Sol-Gel Sci.Tech.* **2001b,** *21, 115.*

17. Smitha, S.; Mukundan, P.; Pillai, P. K.; and Warrier, K. G. K.; *Mater. Chem. Phys.* **2007,** *103,* 318.

18. Pietras, P.; Przekop, R.; and Maciejewski, H.; *Ceramics–Silikáty.* **2013,** *57,* 58–65.

CHAPTER 8

DEVELOPMENT OF COMPUTATIONAL TECHNIQUES IN NANOSYSTEMS

V. I. KODOLOV, N. V. KHOKHRIAKOV, V. V. TRINEEVA,
M. A. CHASHKIN, L. F. AKHMETSHINA, YU. V. PERSHIN, and
YA. A. POLYOTOV

CONTENTS

8.1   THE QUANTUM CHEMICAL INVESTIGATION OF METAL/ CARBON NANOCOMPOSITES INTERACTION WITH DIFFERENT MEDIA AND POLYMERIC COMPOSITIONS

8.1.1   THE QUANTUM CHEMICAL INVESTIGATION OF HYDROXYL FULLERENE INTERACTION WITH WATER

Currently various sources contain experimental data that prove radical changes in the structure of water and other polar liquids when super small quantities of surface active nanoparticles (according to a number of papers—about thousandth of percent by weight) are introduced. When the water modified with nanoparticles is further applied in technological processes, this frequently results in qualitative changes in the properties of products, including the improvement of various mechanical characteristics.

The investigation demonstrates the results of quantum-chemical modeling of similar systems that allow making the conclusion on the degree of nanoparticle influence on the structure of water solutions. The calculations were carried out in the frameworks of ab initio Hartree–Fock method in different basis sets and semi-empirical method PM3. The program complex GAMESS was applied in the calculations [1]. The equilibrium atomic geometries of fragments were defined and the interaction energy of molecular structures E_{int} was evaluated in the frameworks of aforementioned energy models. The interaction energy was calculated by the following formula:

$$E_{int} = E - E_a - E_b,$$ (1)

where E is the energy of the molecular complex with an optimized structure, E_a and E_b are the bond energies of isolated molecules forming the complex. Whencalculating E_a and E_b the molecule energy was not additionally optimized.

At the first stage quantum-chemical investigation of ethyl alcohol molecule interaction with one and a few water molecules was carried out. The calculations were carried out in the frameworks of different semi-empirical and ab initio models. Thus based on the data obtained we can make a conclusion on the error value that can occur due to the use of simplified models with small basis sets. The calculation results are given in Table 8.1.

TABLE 8.1 Equilibrium parameters of the complex formed by the molecules of ethyl alcohol and water

Calculation Method	E_{int}, kcal/mol	R_{9-10}	R_{3-9}	Q_3	Q_9	Q_{10}
PM3	1.88	2.38	0.95	−0.33	0.20	−0.37
3–21G	11.29	1.80	0.97	−0.74	0.41	−0.72
6–31G	7.52	1.89	0.96	−0.81	0.48	−0.83
TZV	6.90	1.91	0.96	−0.64	0.42	−0.83
3–21G MP2	13.17	1.78	1.00	−0.64	0.35	−0.63
TZV**++ MP2	6.27	1.90	0.97	−0.49	0.33	−0.65
TZV MP2	7.52	1.84	0.99	−0.58	0.39	−0.79
TZV+ MP2 for the Geometry Obtained by the Method 6–31G	6.27	1.89	0.96			

The geometrical structure of the complex formed by the ethyl alcohol and water molecules are given in Figure 8.1.

FIGURE 8.1 Equilibrium geometric structure of the complex formed by the molecules of ethyl alcohol and water. optimization by energy is carried out in the basis 6–31G.

The energy minimization of the structure demonstrated was carried out on the basis 6–31G. The atom numbers in Figure 8.1 correspond to the designations given in the Table 8.1.

The calculations demonstrate that equilibrium geometrical parameters of the complex change insignificantly in the calculations by different methods. It should be noted that in the structures obtained without electron correlation the atoms 3, 9, 10, and hydrogen of water molecule are in one plane. The structure planarity is distorted when considering electron correlation by the perturbation theory MP2. Even more considerable deviations from the planarity are observed when calculating by the semi-empirical method PM3. In this case, the straight line connecting the atoms 3 and 9 become practically perpendicular to the water molecule plane. At the same time, the semi-empirical method gives the hydrogen bond length exceeding ab initio results by over 20 percent.

The analysis of energy parameters indicates that the energy of interaction of molecules inside the complex calculated by the Eq. (1) is adequately predicted in the frameworks of Hartree–Fock method in the basis 6–31G. If for the geometrical structure obtained by this method the calculations of energy are made in the frameworks of more accurate models, the calculation results practically coincide with the successive accurate calculations.

The atom charges obtained by different methods vary in a broad range, in this case no regularity is observed. This can be connected with the imperfection of the methodology of charge evaluation by Mulliken.

The investigation of water interaction with the cluster of hydroxyfullerene $C_{60}[OH]_{10}$ was carried out in the frameworks of ab initio Hartree–Fock method in the basis 6–31G. To simplify the calculation model the cluster with the rotation axis of fifth order of fullerene molecule C_{60} was applied. Figure 8.2(a) demonstrates the fullerene molecule structure optimized by energy, and Figure 8.2(b) demonstrates the optimized structure of the cluster $C_{60}[OH]_{10}$.

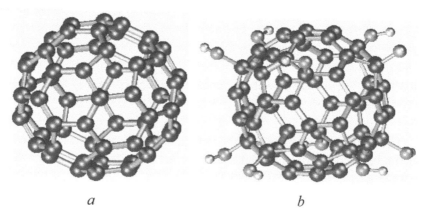

a   *b*

FIGURE 8.2 Equilibrium geometric structure of the molecule C_{60} (Figure 8.2(a)) and cluster $C_{60}[OH]_{10}$ (Figure 8.2(b)). Optimization by energy is carried out in the basis 6–31G.

All the calculations were carried out preserving the symmetry of molecular systems.

The pentagon side length calculated in the fullerene molecule is 1.452 Å, and bond length connecting neighboring pentagons—1.375 Å. These results fit the experimental data available and results of other calculations [2].

To evaluate the energy of interaction of the cluster with water, the complex $C_{60}[OH]_{10} \cdot 10\,H_2O$ with the symmetry C_5 was studied. The complex geometry obtained in the basis 6–31G is given in Figure 8.3.

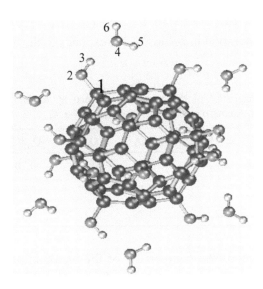

FIGURE 8.3 Molecular system formed by the cluster $C_{60}[OH]_{10}$ and 10 water molecules. Optimization by energy is carried out in the basis 6–31G.

The complex geometrical parameters and atom charges are shown in Table 8.2.

TABLE 8.2   Equilibrium parameters of the complex $C_{60}[OH]_{10} \cdot 10\,H_2O$ (calculations by ab initio method in the basis 6–31G)

Charges					
Q_1	Q_2	Q_3	Q_4	Q_5	Q_6
0.17	−0.78	0.50	−0.86	0.46	0.42
Bond Lengths					
R_{1-2}	R_{2-3}	R_{3-4}	R_{4-5}	R_{4-6}	
1.42	0.97	1.86	0.95	0.95	

The atom numeration in Table 8.2 corresponds to the numeration given in Figure 8.3.

TABLE 8.3   Characteristics of the complexes formed by the molecules $C_{60}[OH]_{10}$, C_2H_5OH with water. Ab initio calculations in the basis 6–31G

	E_{int}, kcal/mol	L,A	Q_H	Q_O (OH)	Q_O (H2O)
$C_{60}(OH)_{10}+10H_2O$	12.69	1.86	0.5	−0.78	−0.86
$C_2H_5OH+H_2O$	7.52	1.89	0.47	−0.81	−0.83
H_2O+H_2O	8.37	1.89	0.42	−0.82	−0.76

Table 8.3 contains the comparison of the breaking off energy of water molecule from the complexes $C_{60}[OH]_{10} \cdot 10\,H_2O$, $C_2H_5OH \cdot H_2O$, $H_2O \cdot H_2O$. Besides, the table gives the lengths of hydrogen bonds in these complexes L and charges of Q atoms of oxygen and hydrogen of OH group and oxygen atom of water molecule. All the calculations are carried out by ab initio Hartree–Fock method in the basis 6–31G. The calculations demonstrate that the energy of interaction of ethyl alcohol molecule with water is somewhat less than the energy of interaction between water molecules. At the same time, nanoparticle $C_{60}[OH]_{10}$ interacts with water molecule about two times more intensively. When comparing the atom charges of the complexes it can be seen that the molecule π—fullerene electrons in the nanoparticle act as a reservoir for electrons. Thus, the molecule OH group is practically electrically neutral. This intensifies the attraction of

water molecule oxygen. Therefore, the strength of hydrogen bond in the complex $C_{60}[OH]_{10} \cdot 10\,H_2O$ increases.

To evaluate the cluster sizes with the center in nanoparticle $C_{60}[OH]_{10}$ that can be formed in water, semi-empirical calculations of energies of attachment of water molecules to the nanoparticle $C_{60}[OH]_2$ with the chain formation are carried out (Figure 8.4). Semi-empirical method PM3 was applied in the calculations. Water molecules are attached to the chain successively; after each molecule is attached, the energy is minimized.

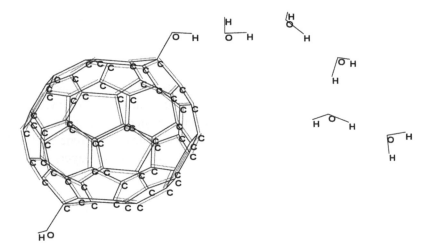

FIGURE 8.4   Model system formed by the cluster $C_{60}[OH]_{10}$ and chain from five water molecules. Optimization by energy is carried out by semi-empirical method PM3.

Table 8.4 contains the results obtained for the chain growth from the nanoparticle, alcohol molecule, and water molecule. Bond energies of water molecules with the chain growing from $C_{60}[OH]_2$ over two times exceed similar energies for the chain growing from OH- group of alcohol molecule and water molecule.

TABLE 8.4   Energies of attachment of water molecule to the chain connected with the molecules $C_{60}[OH]_{10}$, C_2H_5OH. Semi-empirical calculations by the method PM3.

n	$(H_2O)_n$	$C_2H_5OH + n\ H_2O$	$C_{60}(OH)_2 + nH_2O$
1		−1.40	−4.06
2	−1.99	−2.19	−4.83
3	−2.38	−2.51	−5.02
4	−2.54	−2.58	−4.88
5	−2.56	−2.68	−4.71
6	−2.66		−3.04

At the same time, the energy of hydrogen bond first increases till the attachment of the third molecule, and then gradually decreases. When the sixth molecule is attached, the energy decreases in discrete steps.

Thus, the investigations demonstrate that hydroxyfullerene molecule forms a stable complex in water, being surrounded by six layers of water molecules. Obviously, the water structure is reconstructed at considerably larger distances. The introduction of insignificant number of nanoparticles results in complete reconstruction of water medium, and, therefore, in considerable change in the properties of substances obtained on its basis. Thus, the results are reported on the considerable change in the properties of constructional materials obtained on the water basis after the insignificant additions of active graphite-like nanoparticles with chemical properties similar to the ones of hydroxyfullerene.

8.1.2 INFLUENCE COMPARISON OF HYDROXYL FULLERENES AND METAL/CARBON NANOPARTICLES WITH HYDROXYL GROUPS ON WATER STRUCTURE

Quantum-chemical investigations of the interactions between hydroxylated graphite-like carbon nanoparticles, including metal containing ones, and water molecules are given. Hydroxyfullerene molecules, as well as carbon envelopes consisting of hexangular aromatic rings, and envelope containing a defective carbon ring (pentagon or heptagon) are considered as the models of carbon hydroxylated nanoparticles. The calculations demonstrate that hydroxyfullerene molecule $C_{60}(OH)_{10} \cdot (H_2O)_{10}$ forms a stronger hydrogen bond with water molecule than the bond between water molecules. Among the unclosed hydroxylated carbon envelopes, the envelope with a pentagon forms the strongest hydrogen bond with water molecule. The presence of metal nanocluster increases the hydrogen bond energy in 1.5–2.5 times for envelopes consisting of pentagons and hexagons.

At present we can find in the literature experimental data confirming radical changes in the structure of water and other polar liquids when small quantities (about thousandth percent by mass) of surfactants are introduced [3–5]. When further used in technological processes, the water modified by nanoparticles frequently results in qualitative changes of product properties including the improvement of different mechanical characteristics.

The chapter is dedicated to the investigation of water molecule interactions with different systems modeling carbon-enveloped nanoparticles, including those containing clusters of transition metals. All the calculations were carried out following the same scheme. At the first stage, the geometry of complete molecular complex was optimized. At the second stage, the interaction energy of the constituent molecular structures E_{int} was assessed for all optimized molecular complexes. The interaction energy was calculated by the formula $E_{int} = E - E_a - E_b$,

where E is the bond energy of the optimized structure molecular complex, E_a and E_b - are bond energies of isolated molecules forming the complex. When calculating E_a and E_b, the molecule energy was not additionally optimized. The energy of chemical bonds was assessed by the same scheme. All the calculations were carried out with the help of software PC-GAMESS [6]. Both semi-empirical and ab initio methods of quantum chemistry with different basis sets were applied in the calculations. Besides, the method of density functional was used. The more detailed descriptions of calculation techniques are given when describing the corresponding calculation experiments.

At the first stage, the quantum-chemical investigation of the interaction of α-naphthol and β-naphthol molecules with water was carried out within various quantum-chemical models. Thus, based on the data obtained the conclusion on the error value can be made, which can be resulted from the application of simplified methods with insufficient basis sets. There are three isomers of naphthol molecule named α-naphthol (complex with water is shown on Figure 8.1(a)), *trans*-rotamer and *cis*-rotamer of β-naphthol (Figures 8.5(b) and 8.5(c)).

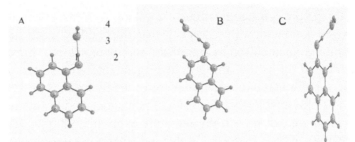

FIGURE 8.5 Equilibrium geometric structure of the complex formed by the molecules of α-naphthol (Figure 8.5(a)), *Trans-β*-naphthol (Figure 8.5(b)) and *Cis-β*-naphthol (Figure 8.5(c)) with water. The optimization by energy was carried out at HF/6–31G Level.

We denote them as α-naphthol, *trans-β*-naphthol, and *cis-β*-naphthol as in what follows. The calculations within all methods applied, except for the semi-empirical one, showed that *cis-β*-naphthol is the most stable isomer, α-naphthol has almost equal stability, and *trans-β*-naphthol is the least stable structure. This conclusion is in accordance with available experimental data [7, 8].

The results for the complexes with α-naphthol are given in Table 8.5.

TABLE 8.5 Equilibrium parameters of the complex formed by α-naphthol molecules and water. In the column E_{int} The interaction energy between water molecules in dimer is given in brackets for comparison

Calcu-lation Method	E_{int}, kJ/mol	r_{1-2}, nm	r_{2-3}, nm	r_{3-4}, nm	q_1	q_2	q_3
PM3	16.27 (8.32)	0.1364	0.0963	0.1819			
HF 3–21G	68.66 (46.11)	0.1361	0.098	0.1701	0.45	–0.82	0.44
HF 6–31G	43.91 (33.13)	0.1365	0.0961	0.1805	0.39	–0.85	0.50
HF 6–31G*	31.91 (23.61)	0.1345	0.0954	0.194	0.42	–0.81	0.52
HF 6–31G* MP2	42.65 (30.75)	0.1367	0.0984	0.1855	0.31	–0.72	0.48
DFT 6–31G* B3LYP	42.58 (32.33)	0.136	0.0983	0.1821	0.28	–0.69	0.44

The geometry structure of the complex formed by naphthols and water molecules is shown in Figure 8.5. The structures given in Figure 8.5 were obtained by minimizing the energy using Hartree–Fock method in the basis 6–31G. The numbers of atoms in Figure 8.5 correspond to those given in Table 8.5. The calculations demonstrate that equilibrium geometric parameters of the complex vary insignificantly when applying different calculation methods. It should be pointed out that in the structures obtained at HF/6–31G level of theory without taking into account of electron correlation, atoms 2, 3, 4, and hydrogens of water molecule are almost in one plane.

The structure planarity is broken with the inclusion of polarization functions into the basis set and electron correlation consideration. The semi-empirical calculation forecasts the planarity disturbance of the atom group considered. The hydrogen bond for β-naphthol complexes is longer than one for α-naphthol. These values are 0.1812 nm for cis-β-naphthol and 0.1819 nm for trans-β-naphthol calculated at HF/6–31G level. The values obtained at MP2/6–31G* level of theory are 0.1861 nm and 0.1862 nm. They are in accordance with the results of more accurate calculations 0.1849 nm and 0.1850 nm [9].

The second column in Table 8.1 demonstrates the energy of interaction between α-naphthol and water molecule calculated using Eq. (1). The interaction energies between water molecules in water dimer are given in brackets for comparison. Although all energy characteristics for different methods vary in a broad range, the ratio of energies E_{int} for the complex of water with α-naphthol and water dimer is between 1.3–1.4. Thus, we can assume that Hartree–Fock method in the basis 6–31G provides qualitatively correct conclusions for the energies of intermolecular interaction.

It should be pointed out that energy E_{int} calculated in the basis 6–31G agrees numerically with the results of more accurate calculations. Nevertheless, the interaction energies between water and α-naphthol exceed experimental value 24.2 ± 0.8 kJ/mol [5]. The energies of interaction between water molecules in dimer are overestimated as well. The errors are known to be caused by the superposition of basis sets of complex molecules (basis superpositional error (BSSE)) and zero-point vibration energy (ZPVE). Calculations completed have shown that BSSE decreases the interaction energy by 10–20 percent depending on the basis, but does not lead to qualitative changes. The ZPVE results in additional decrease in the energies of intermolecular interaction.

Thus taking into account ZPVE and BSSE considerably improves the accordance with the available data. Interaction energy between water and α-naphthol is 30 kJ/mol in the basis 6–31G after correction. In general, analysis of energy parameters shows that all the models considered give the correct correlation for energies and demonstrate that the energy of interaction between water and α-naphthol molecules exceeds the energy of interaction between water molecules. Numerical accordance with experimental value is achieved only after taking into account electronic correlation at the MP2 level.

In this model, the interaction energy after all the corrections mentioned becomes 23.7 kj/mol. *Cis-β*-naphthol is known to be more stable than *trans*-isomer. The energy difference calculated in 6–31G basis set is 3.74 kJ/mol that is equal to experimental value [8]. HF/6–31G binding energy for water dimer is 18.9 kJ/mol after corrections (experimental value is 15±2 kJ/mol [10]). As in the case of α-naphthol complex MP2 calculation is necessary to obtain value that is equal to experimental value.

The vibrational frequencies of naphthols and their complexes with water molecule were calculated and multiplied by a scaling factor of 0.9034 [8]. IR-active frequencies corresponding to naphthol OH-group stretching are 3660 cm^{-1} for *trans-β*-naphthol and 3654 cm^{-1} for *cis-β*-naphthol. They are in accordance with experimental frequencies 3661 cm^{-1} and 3654 cm^{-1} [8]. The influence of hydrogen bond on naphthol OH-group stretching calculated in this chapter is overestimated. Calculated frequencies at HF/6–31G level of theory in the presence of water molecule are 3479 cm^{-1} for *trans-β*-naphthol and 3466 cm^{-1} for *cis-β*-naphthol (for comparing experimental values are 3523 cm^{-1} and 3512 cm^{-1}, respectively). The same result was observed in research [8]. The atom charges obtained within different methods vary in a broad range without any regularity. It can be connected with the imperfectness of Mulliken's charge evaluation technique.

Thus, comparison of calculation results with experimental data demonstrates ab initio Hartree–Fock method in the 6–31G basis set provides qualitatively correct conclusion on hydrogen bond energies for systems considered. This model was applied to investigate the interaction between water and hydroxyfullerene cluster $C_{60}[OH]_{10}$. Hydroxyfullerene is a proper object for quantum-chemical modeling of interaction between activated carbon nanoparticle and water. At the same time hydroxyfullerene is of great interest because of its possible applications in medicine, for water disinfection, and for polishing nanosurfaces.

The cluster with the rotation axis C5 was used in our research of hydroxyfullerene to simplify the calculation model. Figure 8.6(a) demonstrates the structure of C_{60} molecule optimized by energy, and Figure 8.6(b) demonstrates the optimized structure of cluster $C_{60}[OH]_{10}$. All the calculations were carried out conserving the symmetry C_5 of molecular systems.

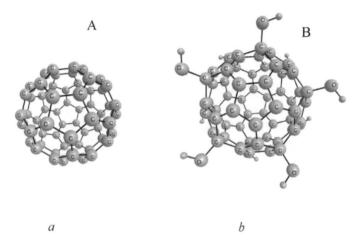

A

B

a   *b*

FIGURE 8.6   Equilibrium geometric structure of C_{60} molecule (*a*) and cluster $C_{60}[OH]_{10}$ (*b*).

The optimization by energy is carried out at HF/6–31G level.

TABLE 8.6   Equilibrium parameters of the complex $C_{60}[OH]_{10} \cdot 10H_2O$ (HF/6–31G calculation)

Charges					
q_1	q_2	q_3	q_4	q_5	q_6
0.17	−0.78	0.50	−0.86	0.46	0.42
Bond lengths, nm					
r_{1-2}	r_{2-3}	r_{3-4}	r_{4-5}	r_{4-6}	
0.142	0.097	0.186	0.095	0.095	

TABLE 8.7 Characteristics of the complexes formed by $C_{60}[OH]_{10}$ and α-naphthol molecules and water HF/ 6–31G calculation.

Complex	E_{int}, kJ/mol	L, nm	q_H	$q_o(OH)$	$q_o(H_2O)$
$C_{60}[OH]_{10} \cdot 10H_2O$	53.09	0.186	0.5	−0.78	−0.86
$C_{10}H_7OH \cdot H_2O$	43.91	0.181	0.5	−0.85	−0.84
$H_2O \cdot H_2O$	33.13	0.189	0.47	−0.88	−0.83

The length of pentagon side in the fullerene molecule was 0.1452 nm, and the length of bond connecting the neighboring pentagons −0.1375 nm. These results agree with the experimental data available and results of other calculations [11, 12].

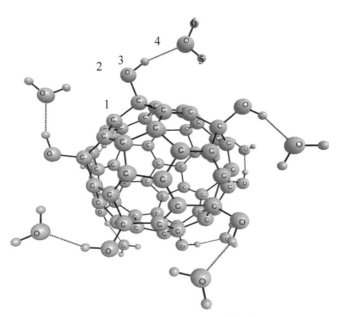

FIGURE 8.7 Molecular system formed by the cluster $C_{60}[OH]_{10}$ and 10 water molecules. The optimization by energy is carried out at HF/ 6–31G Level.

Complex $C_{60}[OH]_{10} \cdot 10H_2O$ with symmetry C_5 was considered to obtain the interaction energy of cluster $C_{60}[OH]_{10}$ and water. The geometric structure of the complex optimized by energy is given in Figure 8.7. The complex geometric parameters and atom charges are given in Table 8.6. The atoms are enumerated in the same way as in Figure 8.7. Table 8.7 contains the water molecule isolation energies from complexes $C_{60}[OH]_{10} \cdot 10H_2O$, $C_{10}H_7OH \cdot H_2O$, $H_2O \cdot H_2O$. Besides, the table contains the lengths of hydrogen bonds in these complexes L and charges Q of oxygen and hydrogen atoms of fullerene OH group and oxygen atom of water molecule. The calculations demonstrate that the energy of interaction between nanoparticle $C_{60}[OH]_{10}$ and water molecule is nearly twice as big as the interaction energy of water molecules in dimer. Basis set superposition error was also analyzed. The calculation indicates that the account of this effect decreases the energies of intermolecular interaction by 15 per cent. The qualitative conclusions from the calculations do not change, and the gain in the interaction energy for the complex $C_{60}[OH]_{10} \cdot 10H_2O$ increases insignificantly in comparison with other complexes. When comparing the atom charges of the complexes, it can be concluded that fullerene π-orbitals in nanoparticle act as a reservoir for electrons. Thus, the charge of hydroxyfullerene OH group decreases. This increases the attraction of water molecule oxygen and the strength of hydrogen bond in complex $C_{60}[OH]_{10} \cdot 10H_2O$.

To estimate the size of the cluster with the center in nanoparticle $C_{60}[OH]_{10}$ that can be formed in water, the semi-empirical calculations of energies of water molecules adding to nanoparticle $C_{60}[OH]_2$ with chain formation were made (Figure 8.8). The semi-empirical method PM3 was used in calculations. Water molecules were bonded to the chain consequently; the energy was minimized after each molecule was added. Table 8.8 demonstrates the comparative results obtained for the chain growth from both nanoparticle and water molecule. The binding energy of water molecule with the chain growing from $C_{60}[OH]_2$ over twice as big as similar energies for the chain growing from the water molecule.

At the same time, the energy of hydrogen bond first increases before the addition of the third molecule and then slowly decreases. When the sixth molecule is added, the energy decreases step-wise. Thus, the research demonstrates that hydroxyfullerene molecule forms a stable complex in water, being surrounded with six layers of water molecules. It is clear that the water structure changes at much longer distances. The introduction of insignificant number of nanoparticles results in complete reconstruction of aqueous medium, and, consequently, to sufficient changes in the properties of substances obtained on its base.

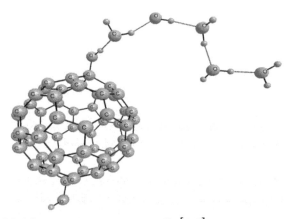

FIGURE 8.8   Model system formed by cluster $C_{60}[OH]_2$ and five water molecule chain. The optimization by energy was carried out with semi-empirical method PM3.

TABLE 8.8   Energies of binding water molecules to the chain connected with molecules $C_{60}[OH]_2$ and water chain (in kJ/mol). Semi-empirical calculations with PM3 method

N	$[H_2O]_n$	$C_{60}[OH]_2 + n \cdot H_2O$
1		16.99
2	8.33	20.21
3	9.96	21.00
4	10.63	20.42
5	10.71	19.71
6	11.13	12.72

In some experimental works carbon nanoparticles are obtained in gels of carbon polymers in the presence of salts of transition metals [13]. At the same time, nanocomposites containing metal nanoparticles in carbon envelopes are formed. The envelope structure contains many defects but it mainly has an aromatic character. The nanoparticles considered are widely used for the modification of different substances; therefore, it is interesting to investigate the influence of metal nuclei on the interaction of carbon-enveloped nanoparticles with polar fluids.

The systems demonstrated in Figures 8.9–8.11 were considered as the nanoparticle model containing a metal nucleus and hydroxylated carbon envelope. In all the figure of the paper, carbon and hydrogen atoms are not signed. At the same time, carbon atoms are marked with dark spheres and hydrogen atoms-with light ones. Clusters comprising two metal atoms are introduced into the system to model a metal nucleus. Carbon envelopes are imitated with 8-nucleus polyaromatic carbon clusters. The carbon envelope indicated in Figure 8.9 contains one pentagonal aromatic ring, and the envelope in Figure 8.11 contains one heptagon. The envelope in Figure 8.10 contains only hexagonal aromatic rings.

The corresponding number of hydrogen atoms is located on the boundaries of carbon clusters. Besides, one OH - group is present in each of the carbon envelopes. Thus, the chemical formula of hydroxylated carbon envelope with pentagonal defect $C_{23}H_{11}OH$ hydroxylated with defectless carbon envelope- $C_{27}H_{13}OH$, and for the hydroxylated carbon envelope with heptagonal defect- $C_{31}H_{15}OH$. Copper and nickel are taken as metals. The interaction of the considered model systems with water molecules is investigated in this chapter.

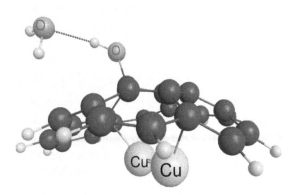

FIGURE 8.9   Complex $Cu_2C_{23}H_{11}OH \cdot H_2O$ with hexagonal defect in carbon envelope.

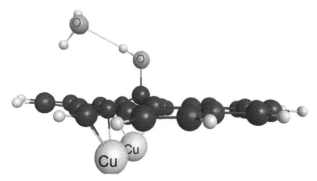

FIGURE 8.10 Complex $Cu_2C_{27}H_{13}OH \cdot H_2O$.

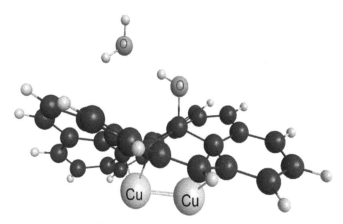

FIGURE 8.11 Complex $Cu_2C_{31}H_{15}OH \cdot H_2O$ with heptagonal defect in carbon envelope.

The calculations are made in the frameworks of density functional with exchange-correlative functional B3LYP. For all the atoms of the model system, apart from oxygen, the basis set 6–31G was applied. Polarization and diffusion functions were added to the basis set for more accurate calculations of hydrogen bond energy and interaction of hydroxyl group with carbon envelope. Thus, the basis set 6–31+G* was used for oxygen atoms. The basis sets from Ref. [14] were used for atoms of transition metals, and the bases from software PC-Games were applied for the rest of the elements.

Hydrogen bond energy in water dimers calculated in this model exceeded 22 kJ/mol being a little more than the value obtained in the frameworks of the more accurate method—19.65 kJ/mol [15]. The experimental value is 14.9 ± 2 kJ/mol [15]. If we consider the energy of zero oscillations of dimer and water molecules in the frameworks of the calculation model, hydrogen bond energy decreases to 12.57 kJ/mol and becomes similar to the experimental value. When the basis extends due to polarization functions on hydrogen atoms, hydrogen bond energy goes down by 2 per cent.

TABLE 8.9 Bond lengths (in angstrems)

Compound	$C-O$	$O-H$	$O \cdots H$
$C_{23}H_{11}OH$	1.492	0.973	–
$Cu_2C_{23}H_{11}OH$	1.450	0.973	–
$Ni_2C_{23}H_{11}OH$	1.454	0.973	–
$C_{27}H_{13}OH$	1.545	0.974	–
$Cu_2C_{27}H_{13}OH$	1.469	0.973	–
$Ni_2C_{27}H_{13}OH$	1.464	0.972	–
$C_{31}H_{15}OH$	1.506	0.973	–
$Cu_2C_{31}H_{15}OH$	1.470	0.973	–
$Ni_2C_{31}H_{15}OH$	1.455	0.972	–
H_2O	–	0.969	–
$C_{23}H_{11}OH \cdot H_2O$	1.470	0.981	1.967
$Cu_2C_{23}H_{11}OH \cdot H_2O$	1.439	0.984	1.869
$Ni_2C_{23}H_{11}OH \cdot H_2O$	1.440	0.985	1.893
$C_{27}H_{13}OH \cdot H_2O$	1.524	0.979	2.119

TABLE 8.9   *(Continued)*

Compound	$C-O$	$O-H$	$O\cdots H$
$Cu_2C_{27}H_{13}OH \cdot H_2O$	1.451	0.984	1.911
$Ni_2C_{27}H_{13}OH \cdot H_2O$	1.449	0.981	1.964
$C_{31}H_{15}OH \cdot H_2O$	1.490	0.979	2.024
$Cu_2C_{31}H_{15}OH \cdot H_2O$	1,453	0.981	1.964
$Ni_2C_{31}H_{15}OH \cdot H_2O$	1.446	0.979	1.997
$H_2O \cdot H_2O$	–	0.978	1.906

At the first stage of the investigation we minimized the energy of all the considered structures in the presence of water molecule and without it. In Figures 8.9–8.11, you can see the optimized geometric structures for complexes containing copper cluster. In the second column of Table 8.5, we can find the lengths of $C-O$ bond formed by OH - group with carbon envelope atom for the calculated complexes, the bond length in hydroxyl group in the third column, and the lengths of hydrogen bond for complexes with a water molecule in the fourth column.

The analysis of geometric parameters demonstrates that in all cases a metal presence results in the decrease of $C-O$ bond length formed by OH - group with the carbon envelope atom. The especially strong effect is observed in the defectless carbon envelope which is explained by a high stability of defectless graphite-like systems. The metal availability results in the increased activity of the carbon envelope due to the additional electron density and destabilization of aromatic rings. The defects of carbon grid have a similar effect.

At the same time, the pentagonal defect is usually characterized by the electron density excess, and the heptagonal one by its lack. If we compare the two metals clusters considered, we see that for the system with pentagonal defect copper influence is stronger than nickel one, and for the rest of the systems we observe the opposite effect. A metal has only an insignificant influence on $O-H$ bond length.

The interaction with water molecule results in the increase of $O-H$ bond length in the hydroxyl group of carbon envelope and decrease of $C-O$ bond length. In the presence of metal the length of hydrogen bond decreases in all the cases, and copper clusters cause the formation of shorter hydrogen bonds. In our calculations, we took into account the BSSE. In Table 8.6 we see the interaction energies of the considered model systems with water molecule. In Table 8.6, the energy of intermolecular interaction in water dimer calculated using the basis 6−31+G* is given for comparison.

From Table 8.6, we observe that the interaction of hydroxylated carbon envelopes with water molecule is significantly weaker than the intermolecular interaction in water dimer. The strongest hydrogen bond is formed in case of the envelope with pentagonal defect $C_{23}H_{11}OH$. This result correlates with the minimal length of $C-O$ bond in the envelope $C_{23}H_{11}OH$ (Table 8.9). Thus, hydrogen bond is getting weaker in the hydroxyl group and it provides the formation of a stronger hydrogen bond. The main conclusion from Table 8.6 is a considerable strengthening of hydrogen bond in metal systems. For complex $Cu_2C_{27}H_{13}OH \cdot H_2O$ with defectless carbon envelope the energy of hydrogen bond increases in 2.5× in comparison with a similar complex without metal. The greatest energy of hydrogen bond is obtained for complex $Ni_2C_{27}H_{13}OH \cdot H_2O$. This energy exceeds the energy of hydrogen bond in water dimer by over 40 per cent.

TABLE 8.10   Interaction energy of model clusters with water molecule (kJ/mol)

Compound	E_{int}
$C_{23}H_{11}OH \cdot H_2O$	18.12
$Cu_2C_{23}H_{11}OH \cdot H_2O$	26.06
$Ni_2C_{23}H_{11}OH \cdot H_2O$	31.56
$C_{27}H_{13}OH \cdot H_2O$	11.01
$Cu_2C_{27}H_{13}OH \cdot H_2O$	27.04

TABLE 8.10   *(Continued)*

Compound	E_{int}
$Ni_2C_{27}H_{13}OH \cdot H_2O$	19.65
$C_{31}H_{15}OH \cdot H_2O$	15.08
$Cu_2C_{31}H_{15}OH \cdot H_2O$	13.75
$Ni_2C_{31}H_{15}OH \cdot H_2O$	16.88
$C_{60}(OH)_2 \cdot H_2O$	21.98
$H_2O \cdot H_2O$	22.37

It should be noted that in the frameworks of model HF/6–31G the energy of hydrogen bond for hydroxyfullerene $C_{60}(OH)_{10} \cdot (H_2O)_{10}$ is higher than the energy of hydrogen bond in water dimer by 60 percent (Table 8.7). The semi-empirical calculations for complex $C_{60}(OH)_2 \cdot H_2O$ predict that the energy of hydrogen bond two times exceeds the one in water dimer. When calculating by the method described in the final part of this chapter, the energy of hydrogen bond for hydroxyfullerene $C_{60}(OH)_{10} \cdot (H_2O)_{10}$ is higher by 37 per cent than the bond energy in water dimer. Thus, taking into account diffusion and polarization functions we considerably change the results. The additional reason for the difference in the calculation of complex $C_{60}(OH)_{10} \cdot (H_2O)_{10}$ is the use of the method of hydrogen bond energy evaluation with breaking off all water molecules while preserving the system symmetry.

The main conclusion from the calculation results presented is a considerable increase in hydrogen bond energy with water molecule for carbon graphite-like nanoparticles containing the nucleus from transition metal atoms. Besides, the energy of $C - O$ bond between carbon envelope and hydroxyl group increases in the presence of metal. Thus, the probability of the formation of hydroxylated nanoparticles increases. The nanoparticles considered have a strong orientating effect on water medium and become the embryos while forming materials with improved characteristics. A similar effect is observed when water molecule interacts with hydroxyfullerene. In this case, the interaction energy depends considerably on the number of OH– groups in the molecule. Thus, the interaction energy of water molecules with hydroxyfullerene in cluster $C_{60}(OH)_{10} \cdot (H_2O)_{10}$ exceeds the interaction energy in water dimer by 40–60 percent depending on the calculation model by 40–60 per cent. At the same time, for complex $C_{60}(OH)_2 \cdot H_2O$ hydrogen bond is a little weaker than in water dimer.

8.1.3 QUANTUM-CHEMICAL INVESTIGATION OF INTERACTION BETWEEN FRAGMENTS OF METAL–CARBON NANOCOMPOSITES AND POLYMERIC MATERIALS FILLED BY METAL CONTAINING PHASE

There is a strong necessity to define formulations components processes conditions for nanostructured materials filled by metallic additives. Another task is optimization of components, nanocomposites and diluents combination and, in what follows, curing processes with determined temperature mode. The result of these arrangements will be materials with layerwise homogeneous metal particles/nanocomposites distribution formulation in ligand shell.

The result of these arrangements will be homogeneous metal particles/ nanocomposites distribution in acetylacetone ligand shell. In the capacity of conductive filler silver nanoclusters/nanocrystals and copper-nickel/ carbon nanocomposites can be used. It should be lead to decrease volume resistance from 10^{-4} to 10^{-6} $\Omega \cdot$cm and increase of adhesive/paste adhesion.

Silver filler is more preferable due to excellent corrosion resistance and conductivity, but its high cost is serious disadvantage. Hence alternative conductive filler, notably nickel-carbon nanocomposite was chosen to further computational simulation. In developed adhesive/paste formulations metal containing phase distribution is determined by competitive coordination and cross-linking reactions.

There are two parallel technological paths that consist of preparing blend based on epoxy resin (ER) and preparing another one based on polyethylene polyamine (PEPA). Then prepared blends are mixed and cured with diluent gradual removing. Curing process can be tuned by complex diluent contains of acetiylacetone (AcAc), diacetonealcohol (DAcA) (in ratio AcAc:DAcA=1:1) and silver particles being treated with this complex diluent. The ER with silver powder mixing can lead to homogeneous formulation formation. Blend based on PEPA formation comprises AA and copper-/nickel-carbon nanocomposites introducing. Complex diluent removing tunes by temperature increasing to 150–160°C. Initial results formulation with 0.01 percent nanocomposite (by introduced metal total weight) curing formation showed that silver particles were self-organized and assembled into layer-chained structures.

Acetylacetone (AcAc) infrared spectra were calculated with Hyper-Chem software. Calculations were carried out by the use of semi-empirical methods PM3 and ZINDO/1 and ab initio method with 6–31G** basis set [16]. Results of calculation have been compared with instrumental measurement performed IR-Fourier spectrometer FSM-1201. Liquid adhesive components were preliminary stirred in ratio 1:1 and 1:2 for the purpose of interaction investigation. Stirring was carried out with magnetic mixer. Silver powder was preliminary crumbled up in agate mortar. Then acetylacetone was added. Obtained mixture was taken for sample after silver powder precipitation.

Sample was placed between two KBr glass plates with identical clamps. In the capacity of comparative sample were used empty glass plates. Every sample spectrum was performed for five times and the final one was calculated by striking an average. Procedure was carried out in transmission mode and then data were recomputed to obtain absorption spectra.

There are four possible states of AA: ketone, ketone-enol, enol A, enol B (Figure 8.12). Vibrational analysis is carried out with one of semi-empirical of ab initio methods and can allow to recognize different states of reagents by spectra comparison.

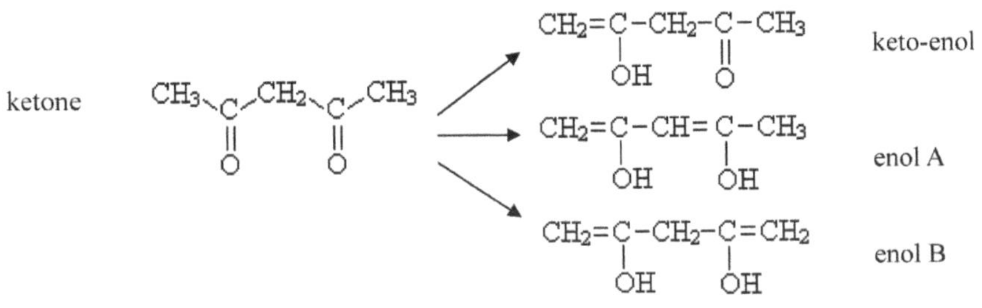

FIGURE 8.12   Different states of acetylacetone.

The vibrational frequencies are derived from the harmonic approximation, which assumes that the potential surface has a quadratic form. The association between transition energy ΔE and frequency v is performed by Einstein's formula

$$\Delta E = h \cdot v \qquad (2)$$

where h is the Planck constant. IR frequencies ($\sim 10^{12}$ Hz) accord with gaps between vibrational energy levels. Thus, each line in an IR spectrum represents an excitation of nuclei from one vibrational state to another [17].

Comparison between MNDO, AM1 and PM3 methods were performed in [18]. Semi-empirical method PM3 demonstrated the closest correspondence to experimental values.

In the Figure 8.13 showed experimental data of IR-Fourier spectrometer AA measurement and data achieved from the National Institute of Advanced Industrial Science and Technology open database. Peaks comparison is clearly shown both spectra generally identical. IR spectra for comparison with calculated ones were received from open AIST database [1]. As it can be concluded from AIST data, experimental IR spectrum was measured for AA ketone state.

As we can see, most intensive peak 1622 cm^{-1} belongs to C=O bond vibrations. Oxygen relating to this bond can participate in different coordination reactions. Peaks 915 cm^{-1} and 956cm^{-1} belong to O-H bond vibrations. This bond can also take part in many reactions.

Computational part is comprised several steps. There were several steps in computational part. Firstly, geometry of AA was optimized to achieve molecule stable state. Then vibrational analysis was carried out. Parameters of the same computational method were used for subsequent IR spectrum calculation. Obtained IR spectra are shown in Figure 8.14.

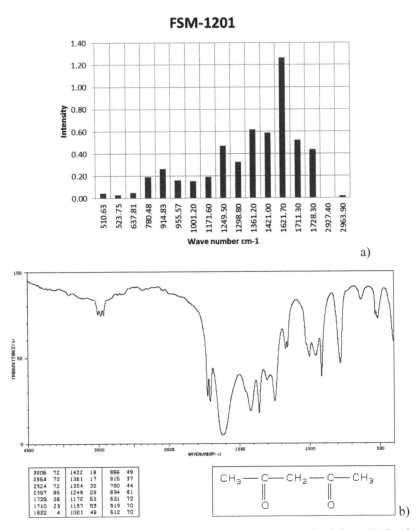

a)

b)

FIGURE 8.13 Experimental IR spectrum of acetylacetone obtained from IR-Fourier FSM-1201 spectrometer (a) and IR spectrum of acetylacetone received from open AIST database (b).

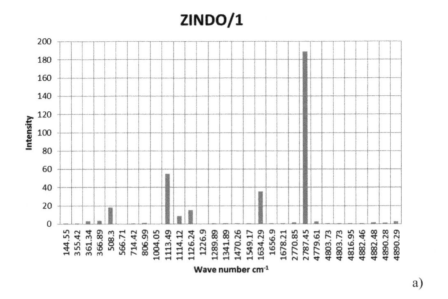

a)

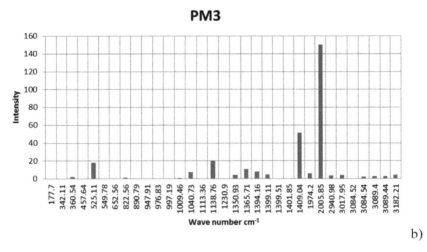

b)

FIGURE 8.14   *(Continued)*

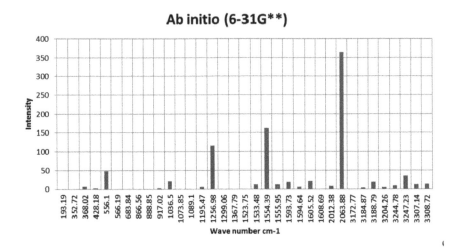

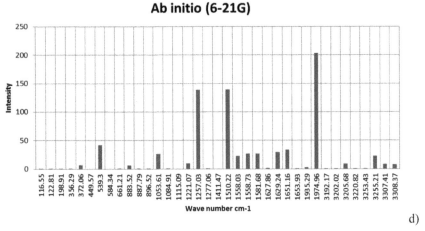

d)

FIGURE 8.14 IR spectra of acetylacetone calculated with hyperChem software: semi-empirical method ZINDO/1 (a); semi-empirical method PM3 (b); ab initio method with 6–31G** basis set (c); ab initio method with 6–21G basis set (d).

Vibrational analysis performed by HyperChem was showed that PM3 demonstrates more close correspondence with experimental and ab initio calculated spectra than ZINDO/1. As expected, ab initio spectra were demonstrated closest result in general. As we can see at Figure 8.14(c) and 8.14(d), peaks of IR spectrum calculated with 6–21G small basis set are staying closer with respect to experimental values than ones calculated with large basis set 6–31G**. Thus it will be reasonable to use both PM3 and ab initio with 6–21G basis set methods in further calculations.

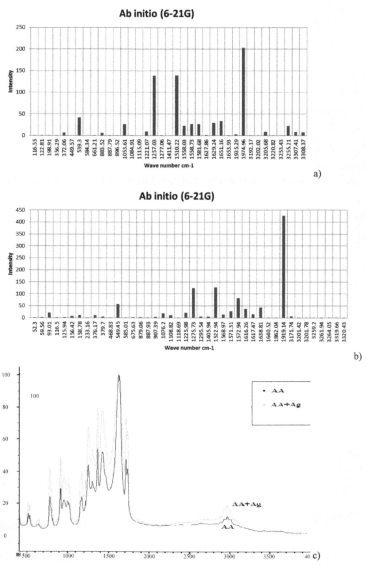

FIGURE 8.15 IR spectra obtained with ab initio method with 6–21G basis set: acetylacetone (a); acetylacetone with silver powder (b); experimental data obtained from IR-Fourier FSM-1201 spectrometer: acetylacetone with silver powder in comparison with pure acetylacetone (c).

As shown on Figures 8.15(a) and 8.15(b), peaks are situated almost identical due to each other with the exception of peak 1,573 cm^{-1} (Figure 8.15(b)) which was appeared after silver powder addition. Peak 1,919 cm^{-1} (Figure 8.15(b)) has been more intensive as compared with similar one on Figure 8.15(a).

Peaks are found in range 600–700 cm^{-1} should usually relate to metal complexes with acetylacetone [19, 20]. IR spectra AA and AA with Ag$^+$ ion calculations were carried out by ab initio method with 6–21G basis set. PM3 wasn't used due this method hasn't necessary parametrization for silver. In the case of IR spectra ab initio computation unavoidable calculating error is occurred hence all peaks have some displacement. Peak 549 cm^{-1} (Figure 8.15(b)) is equivalent to peak 638 cm^{-1} (Figure 8.15(c)) obtained experimentally and it is more intensive than similar one on Figure 8.4. Peak 1974 cm^{-1} was displaced to mark 1919 cm^{-1} and became far intensive. It can be explained Ag+ influence and coordination bonds between metal and AA formation.

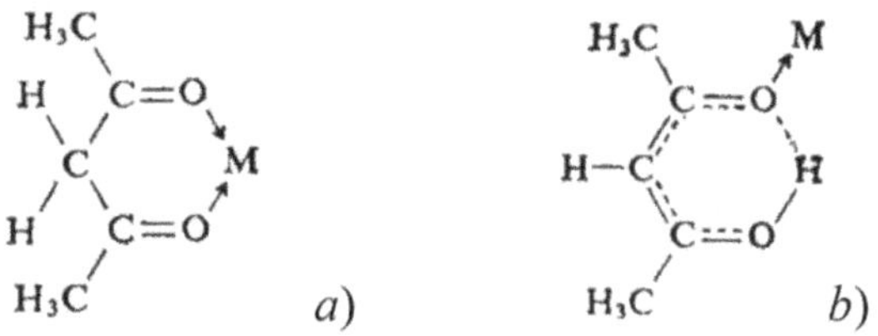

FIGURE 8.16   Different ways of complex between metal and AA formation: kentone with metal (a); enol with metal (b).

Complexes between silver and AA depending on enol or ketone state can form with two different ways (Figure 8.16).

IR spectra of pure acetylacetone and acetylacetone with silver were calculated with different methods of computational chemistry. PM3 method was shown closest results with respect to experimental data among semi-empirical methods. It was established that there is no strong dependence between size of ab initio method basis set and data accuracy. It was found that IR spectrum carried out with ab initio method with small basis set 6–21G has more accurate data than analogous one with large basis set 6–31G**. Experimental and calculated spectra were shown identical picture of spectral changes with except of unavoidable calculating error.

Hence PM3 and ab initio methods can be used in metal-carbon complexes formations investigations.

8.2   THE METAL/CARBON NANOCOMPOSITES INFLUENCE MECHANISMS ON MEDIA AND ON COMPOSITIONS

The modification of materials by nanostructures (NS) including metal/carbon nanocomposites consists in the conducting of the following stages:

1.   The choice of liquid phase for the making of finely dispersed suspensions intended for the definite material or composition.
2.   The making of finely dispersed suspension with sufficient stability for the definite composition.
3.   The development of conditions of the finely dispersed suspension introduction into composition.

At the choice of liquid phase for the making of finely dispersed suspension it should be taken the properties of NS (nanocomposites) as well as liquid phase (polarity, dielectric penetration, viscosity).

It is perspective if the liquid phase completely enter into the structure of material formed during the composition hardening process. When the correspondent solvent, on the base of which the suspension is obtained, is evaporated, the re-coordination of nanocomposite on other components takes place and the effectiveness of action of NS on composition is decreased.

The stability of finely dispersed suspension is determined on the optical density. The time of the suspension optical density conservation defines the stability of suspension. The activity of suspension is found on the bands intensity changes by means of IR and Raman spectra. The intensity increasing testify to transfer of NS surface energy vibration part on the molecules of medium or composition. The line speading in spectra testify to the growth of electron action of nanocomposites with medium molecules. Last fact is confirmed by x-ray photoelectron investigations.

The changes of character of distribution on nanoparticles sizes take place depending on the nature of nanocomposites, dielectric penetration and polarity of liquid phase. Below characteristics of finely dispersed suspensions of metal/carbon nanocomposites are given. The distribution of nanoparticles in water, alcohol and water-alcohol suspensions prepared based on the above technique are determined with the help of laser analyzer. In Figures 8.17, 8.18 you can see distributions of copper/carbon nanocomposite in the media different polarity and dielectric penetration.

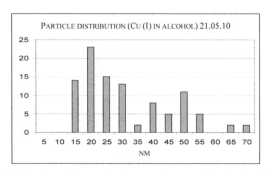

FIGURE 8.17   Distribution of copper/carbon nanocomposites in alcohol.

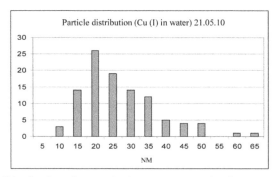

FIGURE 8.18   Distribution of copper/carbon nanocomposites in water.

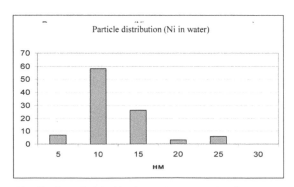

FIGURE 8.19   Distribution of nickel/carbon nanocomposites in water.

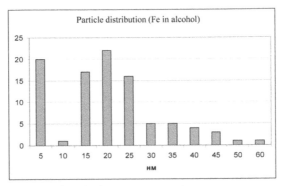

FIGURE 8.20   Distribution of iron/carbon nanocomposites.

When comparing the figure we can see that ultrasound dispergation of one and the same nanocomposite in media different by polarity results in the changes of distribution of its particles. In water solution the average size of Cu/C nanocomposite equals 20 nm, and in alcohol medium—greater by 5 nm.

Assuming that the nanocomposites obtained can be considered as oscillators transferring their oscillations onto the medium molecules, we can determine to what extent the IR spectrum of liquid medium will change, e.g. polyethylene polyamine applied as a hardener in some polymeric compositions, when we introduce small and supersmall quantities of nanocomposite into it.

IR spectra demonstrate the change in the intensity at the introduction of metal/carbon nanocomposite in comparison with the pure medium (IR spectra are given in Figure 8.21). The intensities of IR absorption bands are directly connected with the polarization of chemical bonds at the change in their length, valence angles at the deformation oscillations, i.e. at the change in molecule normal coordinates.

When NS are introduced into media, we observe the changes in the area and intensity of bands, that indicates the coordination interactions and influence of NS onto the medium (Figures 8.21–8.23).

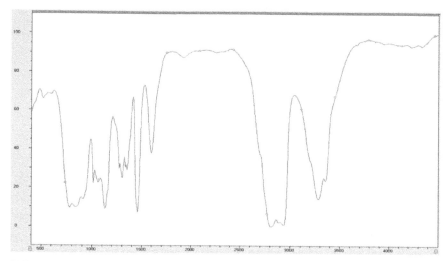

FIGURE 8.21 IR spectrum of polyethylene polyamine.

Special attention in PEPA spectrum should be paid to the peak at 1598 cm^{-1} attributed to deformation oscillations of N-H bond, where hydrogen can participate in different coordination and exchange reactions.

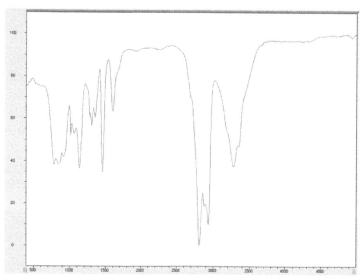

FIGURE 8.22 IR spectrum of copper/carbon nanocomposite finely dispersed suspension in polyethylene polyamine medium (ω (NC) = 1%).

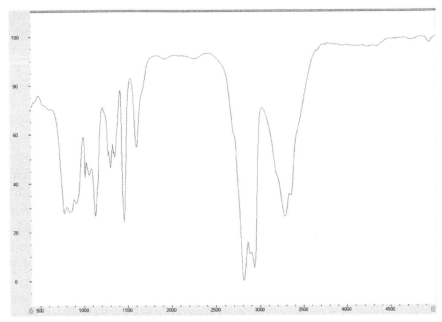

FIGURE 8.23 IR spectrum of nickel/carbon nanocomposite fine suspension (FS) in polyethylene polyamine.

$$(\omega \, (NC) = 1\%)$$

In the spectra wave numbers characteristic for symmetric vs(NH$_2$) 3352 cm^{-1} and asymmetric vas(NH$_2$) 3280 cm^{-1} oscillations of amine groups are present. There is a number of wave numbers attributed to symmetric vs(CH$_2$) 2933 cm^{-1} and asymmetric valence vas(CH2) 2803 cm^{-1}, deformation wagging oscillations vD(CH$_2$) 1349 cm^{-1} of methylene groups, deformation oscillations of NH vD (NH) 1596 cm^{-1} and NH$_2$ vD(NH$_2$) 1456 cm^{-1} amine groups. The oscillations of skeleton bonds at v(CN) 1,059–1,273 cm^{-1} and v(CC) 837 cm^{-1} are the most vivid. The analysis of intensities of IR spectra of PEPA and fine suspensions (FS) of metal/carbon nanocomposites based on it revealed a significant change in the intensities of amine groups of dispersion medium (for vs(NH$_2$) in 1.26 times, and for vas(NH$_2$) in approximately 50 times).

Such demonstrations are presumably connected with the distribution of the influence of nanoparticle oscillations onto the medium with further structuring and stabilizing of the system. Under the influence of nanoparticle the medium changes which is confirmed by the results of IR spectroscopy (change in the intensity of absorption bands in IR region). Density, dielectric penetration, viscosity of the medium are the determining parameters for obtaining FS with uniform distribution of particles in the volume. At the same time, the structuring rate and consequently the stabilization of the system directly depend on the distribution by particle sizes in suspension.

At the wide range of particle distribution by sizes, the oscillation frequency of particles different by size can significantly differ, in this connection, the distortion in the influence transfer of nanoparticle system onto the medium is possible (change in the medium from the part of some particles can be balanced by the other). At the narrow range of nanoparticle distribution by sizes the system structuring and stabilization are possible. With further adjustment of the components such processes will positively influence the processes of structuring and self-organization of final composite system determining physical-mechanical characteristics of hardened or hard composite system.

The effects of the influence of NS at their interaction into liquid medium depend on the type of NS, their content in the medium and medium nature. Depending on the material modified, FS of NS based on different media are used. Water and water solutions of surface-active substances, plasticizers, foaming agents (when modifying foam concretes) are applied as such media to modify silicate, gypsum, cement and concrete compositions. To modify epoxy compounds and glues based on ERs the media based on polyethylene polyamine, isomethyltetrahydrophthalic anhydride, toluene and alcohol—acetone solutions are applied.

To modify polycarbonates and derivatives of polymethyl methacrylate dichloroethane and dichloromethane media are used. To modify polyvinyl chloride compositions and compositions based on phenolformaldehyde and phenolrubber polymers alcohol or acetone-based media are applied. The FS of metal/carbon nanocomposites are produced using the above media for specific compositions. In IR spectra of all studied suspensions the significant change in the absorption intensity, especially in the regions of wave numbers close to the corresponding nanocomposite oscillations, is observed. At the same time, it is found that the effects of nanocomposite influence on liquid media (FS) decreases with time and the activity of the corresponding suspensions drops.

The time period in which the appropriate activity of nanocomposites is kept changes in the interval 24 hours – 1 month depending on the nanocomposite type and nature of the basic medium (liquid phase in which nanocomposites dispergate). For instance, IR spectroscopic investigation of FS based on isomethyltetrahydrophthalic anhydride containing 0.001% of Cu/C nanocomposite indicates the decrease in the peak intensity, which sharply increased on the third day when nanocomposite was introduced (Figure 8.24). Similar changes in IR spectra take place in water suspensions of metal/carbon nanocomposites based on water solutions of surface-active nanocomposites.

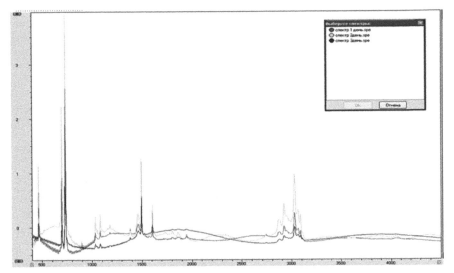

FIGURE 8.24   Changes in IR spectrum of copper/carbon nanocomposite FS based on isomethyltetrahydrophthalic anhydride with time.

(a – IR Spectrum on the First Day After the Nanocomposite was Introduced, b – IR Spectrum on the Second Day, c – IR Spectrum on the Third Day)

In Figure 8.25 you can see IR spectrum of iron/carbon nanocomposite based on water solution of sodium lignosulfonate in comparison with IR spectrum of water solution of surface-active substance.

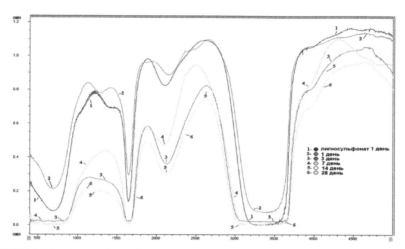

FIGURE 8.25 Comparison of IR spectra of water solution of sodium lignosulfonate (1) and FS of iron/carbon nanocomposite (0.001%) based on this solution on the first day after nanocomposite introduction (2), on the third day (3), on the seventh day (4), 14th day (5) and 28th day. (6)

As it is seen, when nanocomposite is introduced and undergoes ultrasound dispergation, the band intensity in the spectrum increases significantly. Also the shift of the bands in the regions $1100 - 1,300$ cm^{-1}, $2,100 - 2,200$ cm^{-1} is observed, which can indicate the interaction between sodium lignosulfonate and nanocomposite. However after 2 weeks the decrease in band intensity is seen. As the suspension stability evaluated by the optic density is 30 days, the nanocomposite activity is quite high in the period when IR spectra are taken. It can be expected that the effect of foam concrete modification with such suspension will be revealed if only 0.001% of nanocomposite is introduced.

8.3 THE COMPOSITIONS MODIFICATION PROCESSES BY METAL/CARBON NANOCOMPOSITES

8.3.1 GENERAL FUNDAMENTALS OF POLYMERIC MATERIALS MODIFICATION BY METAL/CARBON NANOCOMPOSITES

The material modification with the using of Metal/Carbon Nanocomposites is usually carried out by finely dispersed suspensions containing solvents or components of polymeric compositions. We realize the modification of the following materials: concrete foam, dense concrete, water glass, polyvinyl acetate, polyvinyl alcohol, polyvinyl chloride, polymethyl methacrylate, polycarbonate, ERs, phenol-formaldehyde resins, reinforced plastics, glues, pastes, including current conducting polymeric materials and filled polymeric materials.

Therefore it is necessary to use the different finely dispersed suspension for the modification of enumerated materials. The series of suspensions consist the suspensions on the basis of following liquids: water, ethanol, acetone, benzene, toluene, dichlorethane, methylene chloride, oleic acid, polyethylene polyamine, isomethyl tetra hydrophtalic anhydrite, water solutions surface-active substances or plasticizers. In some cases the solutions of correspondent polymers are applied for the making of the stable finely dispersed suspensions. The estimation of suspensions stability is given as the change of optical density during the definite time (Figure 8.26)

FIGURE 8.26 The change of optical density of typical suspension depending on time.

Relative change of free energy of coagulation process also may be as the estimation of suspension stability –

$$\Delta\Delta F = \lg k_{NC}/k_{NT} \qquad (3),$$

where $k_{NC} - \exp(-\Delta F_{NC}/RT)$ – the constant of coagulation rate of nano-composite, $k_{NT} - \exp(-\Delta F_{NT}/RT)$ – the constant of coagulation rate of carbon nanotube, ΔF_{NC}, ΔF_{NT} – the changes of free energies of corresponding systems (Figure 8.27).

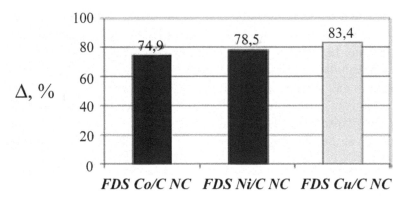

FIGURE 8.27   The comparison of finely dispersed suspensions stability.

The stability of metal/carbon nanocomposites suspension depend on the interactions of solvents with nanocomposites participation (Figure 8.28).

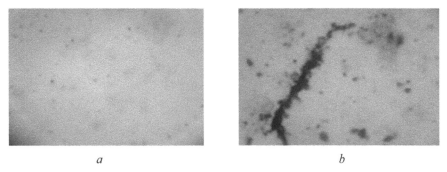

a   *b*

FIGURE 8.28   Microphotographs of Co/C nanocomposite suspension on the basis of mixture "dichlorethane—oleic acid" (*a*) and Oleic Acid (*b*).

The introduction of metal/carbon nanocomposites leads to the changes of kinematic and dynamic viscosity (Figure 8.29).

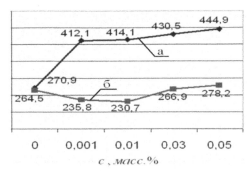

FIGURE 8.29 The dependence of dynamic and kinematic viscosity on the Cu/C nanocomposite quantity.

For the description of polymeric composition self organization the critical parameters (critical content of NS critical time of process realization, critical energetic action) may be used. The equation of NS influence on medium is proposed –

$$W = n/N \exp\{an\tau^\beta/T\} \tag{4}$$

where n – number of active NS, N – number of NS interaction, a – activity of NS, τ - duration of self organization process, T – temperature, β - degree of freedom (number of process direction).

8.3.2 ABOUT THE MODIFICATION OF FOAM CONCRETE BY METAL/CARBON NANOCOMPOSITES SUSPENSION

The ultimate breaking stresses were compared in the process of compression of foam concretes modified with copper/carbon nanocomposites obtained in different nanoreactors of polyvinyl alcohol [21, 22]. The sizes of nanoreactors change depending on the crystallinity and correlation of acetate and hydroxyl groups in PVA which results in the change of sizes and activity of nanocomposites obtained in nanoreactors. It is observed that the sizes of nanocomposites obtained in nanoreactors of PVA matrixes 16/1 (ros) (NC2), PVA 16/1 (imp) (NC1), PVA 98/10 (NC3), correlate as NC3 > NC2 > NC1. The smaller the nanoparticle size the greater its activity, and the less amount of NS is required for self-organization effect.

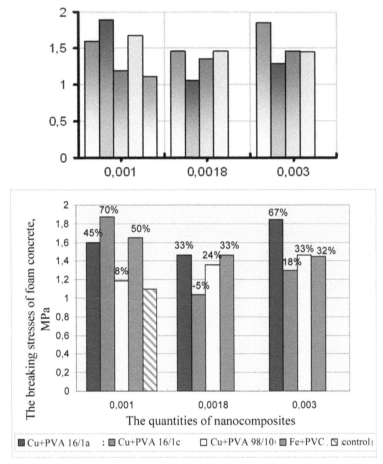

FIGURE 8.30 The dependence of breaking stresses on quantity (%) of metal/carbon nanocomposites.

At the same time, the oscillatory nature of the influence of these nano-composites on the compositions of foam concretes is seen in the fact that if the amount of nanocomposite is 0.0018% from the cement mass, the significant decrease in the strength of NC1 and NC2 is observed. The increase in foam concrete strength after the modification with iron/carbon nanocomposite is a little smaller in comparison with the effects after the application of NC1 and NC2 as modifiers. The corresponding effects after the modification of cement, silicate, gypsum, and concrete compositions with NS is defined by the features of components and technologies applied.

These features often explain the instability of the results after the modification of the foregoing compositions with NS. Besides during the modification the changes in the activity of FS of NS depending on the duration and storage conditions should be taken into account. In this regard, it is advisable to use metal/carbon nanocomposites when modifying polymeric materials whose technology was checked on strictly controlled components.

In this chapter the possibilities of developing new ideas about self-organization processes and about NS and nanosystems are discussed on the example of metal/carbon nanocomposites. The obtaining of metal/carbon nanocomposites in nanoreactors of polymeric matrixes is proposed to consider as self-organization process similar to the formation of ordered phases which can be described with Avrami equation. The application of Avrami equations during the synthesis of nanofilm structures containing copper clusters has been tested. The influence of NS on active media is given as the transfer of oscillation energy of the corresponding NS onto the medium molecules.

The IR spectra of metal/carbon and their finely dispersed suspensions in different media (water and organic substances) have been studied for the first time. It has been found that the introduction of super small quantities of prepared nanocomposites leads to the significant change in band intensity in IR spectra of the media. The attenuation of oscillations generated by the introduction of nanocomposites after the time interval specific for the pair "nanocomposite—medium" has been registered.

Thus, to modify compositions with finely dispersed suspensions it is necessary for the latter to be active enough that should be controlled with IR spectroscopy. A number of results of material modification with finely dispersed suspensions of metal/carbon nanocomposites are given, as well as the examples of changes in the properties of modified materials based on concrete compositions, epoxy and phenol resins, polyvinyl chloride, polycarbonate, and current-conducting polymeric materials.

8.3.3 THE MODIFICATION OF EPOXY RESINS BY METAL/CARBON NANOCOMPOSITES SUSPENSIONS

The IR spectra of polyethylene polyamine and metal/carbon nanocomposite FS based on it with concentrations 0.001%–0.03% from polyethylene polyamine mass are analyzed and their comparative analysis is presented. The possible processes flowing in FS during the interaction of copper/carbon nanocomposite and polyethylene polyamine are described, as well as the processes influencing the increase in adhesive strength and thermal stability of metal/carbon nanocomposite/epoxy compositions.

This decade has been heralded by a large-scale replacement of conventional metal structures with structures from polymeric composite materials (PCM). Currently, we are facing the tendency of production growth of PCM with improved operational characteristics. In practice, PCM characteristics can be improved when applying modern manufacturing technologies, e.g., the application of "binary" technologies of prepreg production [23], as well as the synthesis of new polymeric PCM matrixes or modification of the existing polymeric matrixes with different fillers.

The most cost-efficient way to improve operational characteristics is to modify the existing polymeric matrixes; therefore, currently the group of polymeric materials modified with NS is of special interest. The NS are able to influence the supermolecular structure, stimulate self-organization processes in polymeric matrixes in supersmall quantities, and thus contributing to the efficient formation of a new phase "medium modified – nanocomposite" and qualitative improvement of the characteristics of final product – PCM. This effect is especially visible when NS activity increases which directly depends on the size of specific surface, shape of the particle, and its ultimate composition [22]. Metal ions in the NS used in this work also contribute to the activity increase as they stimulate the formation of new bonds.

The increase in the attraction force of two particles is directly proportional to the growth of their elongated surfaces; therefore, the possibility of NS coagulation and decrease in their efficiency as modifiers increases together with their activity growth. This condition and the fact that the effective concentrations to modify polymeric matrixes are usually in the range <0.01 mass percent impose specific methods for introducing NS into the material modified. The most justified and widely applicable are the methods for introducing NS with the help of FS.

Quantum-chemical modeling methods allow quite precisely defining the typical reaction of interaction between the system components, predict the properties of molecular systems, and decrease the number of experiments due to the imitation of technological processes. The computational results are completely comparable with the experimental modeling results.

The optimal composition of nanocomposite was found with the help of software HyperChem v. 6.03. Cu/C nanocomposite is the most effective for modifying the ER. Nanosystems formed with this NC have higher interaction energy in comparison with nanosystems produced with Ni/C and Co/C nanocomposites. The effective charges and geometries of nanosystems were found with semi-empirical methods.

The fact of producing stable FS of nanocomposite on PEPA basis and increasing the operational characteristics of epoxy polymer was ascertained. It was demonstrated that the introduction of Cu/C nanocomposite into PEPA facilitates the formation of coordination bonds with nitrogen of amine groups, thus resulting in PEPA activity increase in ER hardening reactions. It was found out that the optimal time period of ultrasound processing of copper/carbon nanocomposite FS is 20 min.

The dependence of Cu/C nanocomposite influence on PEPA viscosity in the concentration range 0.001–0.03% was found. The growth of specific surface of NC particles contributes to partial decrease in PEPA kinematic viscosity at concentrations 0.001% and 0.01% with its further elevation at the concentration 0.03%. The IR investigation of Cu/C nanocomposite FS confirms the quantum-chemical computational experiment regarding the availability of NC interactions with PEPA amine groups. The intensity of these groups increased several times when Cu/C nanocomposite was introduced.

The test for defining the adhesive strength and thermal stability correlate with the data of quantum-chemical calculations and indicate the formation of a new phase facilitating the growth of cross-links number in polymer grid when the concentration of Cu/C nanocomposite goes up. The optimal concentration for elevating the modified ER adhesion equals 0.003% from ER weight.

At this concentration the strength growth is 26.8%. From the concentration range studied, the concentration 0.05% from ER weight is optimal to reach a high thermal stability. At this concentration the temperature of thermal destruction beginning increases up to 195°C. Thus, in this work the stable FS of Cu/C nanocomposite were obtained. The modified polymers with increased adhesive strength (by 26.8%) and thermal stability (by 110°C) were produced based on epoxy resins and FS.

8.3.4 THE MODIFICATION OF GLUES BASED ON PHENOL–FORMALDEHYDE RESINS BY METAL/CARBON NANOCOMPOSITES FINELY DISPERSED SUSPENSION

In modern civil and industrial engineering, mechanical engineering, and so on, extra strong, rather light, durable metal structures, wooden structures, and light assembly structures are widely used. However, they deform lose stability and load-carrying capacity under the action of high temperatures; therefore, fire protection of such structures is important issue for investigation.

To decrease the flammability of polymeric coatings it is advisable to create and forecast the properties of materials with external intumescent coating containing active structure-forming agents —regulators of foam cokes structure. At present, metal/carbon nanocomposites are such perspective modifiers. The introduction of nanocomposites into the coating can improve the material behavior during the combustion, retarding the combustion process [24, 25]. In this regard, it is advisable to modify metal/carbon nanocomposites with ammonium phosphates to increase the compatibility with the corresponding compositions.

The grafting of additional functional groups is appropriate in strengthening the additive interaction with the matrix and, thus, in improving the material properties. The grafting can contribute to improving the homogeneity of NS distribution in the matrix and suspension stability. Owing to partial ionization additional groups produce a small surface charge with the result of NS repulsion from each other and stabilization of their dispersion. Phosphorylation leads to the increase in the influence of NS on the medium and material being modified, as well as to the improvement of the quality of NS due to metal reduction [26–28].

The glue BF-19 (based on phenol–formaldehyde resins) is intended for gluing metals, ceramics, glass, wood, and fabric in hot condition, as well as for assembly gluing of cardboard, plastics, leather, and fabrics in cold condition. The glue compositions are organic solvent, synthetic resin (phenol–formaldehyde resins of new lacquer type), and synthetic rubber. When modifying the glue composition, at the first stage the mixture of alcohol suspension (ethyl alcohol + Me/C NC modified) and ammonium polyphosphate (APPh) was prepared. At the same time, the mixtures containing ethyl alcohol, Me/C NC, and APPh, were prepared. At the second stage the glue composition was modified by the introduction of phosphorus containing compositions prepared into the glue BF-19.

The interaction of nanocomposites with APPh results in grating phosphoryl groups to them that allowed using these nanocomposites to modify intumescent fireproof coatings. The sorption ability of functionalized nanocomposites was studied. It was found out that phospholyrated NS have higher sorption ability than nanocomposites not containing phosphorus. Since the sorption ability indicates the nanoproduct activity degree, it can be concluded that nanocomposite activity increases in the presence of active medium and modifier (APPh). The IR spectra of suspensions of nanocomposites on alcohol basis were investigated. It was found that changes in IR spectrum by absorption intensity indicate the presence of excitation source in the suspension. It was assumed that NS are such source of band intensity increase in suspension IR spectra.

The glue coatings were modified with metal/carbon nanocomposite. It was determined that nanocomposite introduction into the glue significantly decreases the material flammability. The samples with phospholyrated nanocomposites have better test results. When phosphorus containing nanocomposite is introduced into the glue, foam coke is formed on the sample surface during the fire exposure. The coating flaking off after flame exposure was not observed as the coating preserved good adhesive properties even after the flammability test.

The nanocomposite surface phospholyration allows improving the nanocomposite structure, increases their activity in different liquid media thus increasing their influence on the material modified. The modification of coatings with nanocomposites obtained finally results in improving their fire-resistance and physical and chemical characteristics.

8.3.5 THE MODIFICATION OF POLYCARBONATE BY METAL/CARBON NANOCOMPOSITES

Recently, the materials based on polycarbonate have been modified to improve their thermal-physical and optical characteristics and to apply new properties to use them for special purposes. The introduction of NS into materials facilitates self-organizing processes in them. These processes depend on surface energy of NS which is connected with energy of their interaction with the surroundings. It is known [21] that the surface energy and activity of nanoparticles increase when their sizes decrease.

For nanoparticles the surface and volume are defined by the defectiveness and form of conformation changes of film NS depending on their crystallinity degree. However, the possibilities of changes in nanofilm shapes with the changes in medium activity are greater in comparison with NS already formed. At the same time, sizes of nanofilms formed and their defectiveness, i.e. tears and cracks on the surface of nanofilms play an important role [22]

When studying the influence of supersmall quantities of substances introduced into polymers and considerably changing their properties, apparently we should consider the role which these substances play in polymers possessing highly organized supermolecular regularity both in crystalline and in amorphous states. It can be assumed that the mechanism of this phenomenon is in the NS energy transfer to polymer structural formations through the interface resulting in the changes in their surface energy and mobility of structural elements of the polymeric body. Such mechanism is quite realistic as polymers are structural-heterogenic (highly dispersed) systems.

Polycarbonate modification with supersmall quantities of Cu/C nanocomposite is possible using FS of this nanocomposite which contributes to uniform distribution of nanoparticles in polycarbonate solution. Polycarbonate "Actual" was used as the modified polycarbonate. The FS of copper/carbon nanocomposite was prepared combining 1.0, 0.1, 0.01, and 0.001% of nanocomposite in polycarbonate solution in ethylene dichloride. The suspensions underwent ultrasonic processing.

To compare the optical density of nanocomposite suspension in polycarbonate solution in ethylene dichloride, as well as polycarbonate and polycarbonate samples modified with nanocomposites, spectrophotometer KFK-3-01a was used. Samples in the form of modified and non-modified films for studying IR spectra were prepared precipitating them from suspension or solution under vacuum. The obtained films about 100 mcm thick were examined on Fourier-spectrometer FSM 1201.To investigate the crystallization and structures formed the high-resolution microscope (up to 10 mcm) was used.

To examine thermal–physical characteristics the lamellar material on polycarbonate and polymethyl methacrylate basis about 10 mm high was prepared. Three layers of polymethyl methacrylate and two layers of polycarbonate were used. Thermal–physical characteristics (specific thermal capacity and thermal conductivity) were investigated on calorimeters IT-c-400 and IT-λ-400.During the investigation the nanocomposite FS in polycarbonate solution in ethylene dichloride was studied, polycarbonate films modified with different concentrations of nanocomposite were compared with the help of optical spectroscopy, microscopy, IR spectroscopy, and thermal–physical methods of investigation.

The results of investigation of optical density of Cu/C nanocomposite FS (0.001%) based on polycarbonate solution in ethylene dichloride are given in Figure 8.31. As seen in Figure 8.31, the introduction of nanocomposite and polycarbonate into ethylene dichloride resulted in transmission increase in the range 640–690 nm (approximately in three times). At the same time, the significant increase in optical density was observed at 790 and 890 nm. The comparison of optical densities of films of polycarbonate and modified materials after the introduction of different quantities of nanocomposite (1.0, 0.1, 0.01%) into polycarbonate is interesting (Figure 8.32).

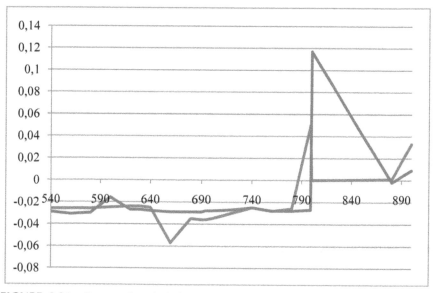

FIGURE 8.31 Curves of optical density of suspension based on ethylene dichloride diluted with polycarbonate and Cu/C nanocomposite in the concentration to polycarbonate 0.001% (1) and ethylene dichloride (2).

Comparison of optical density of suspension containing 0.001% of nanocomposite and optical density polycarbonate sample modified with 0.01% of nanocomposite indicates the proximity of curves character. Thus, the correlation of optical properties of suspensions of nanocomposites and film materials modified with the same nanocomposites is quite possible.

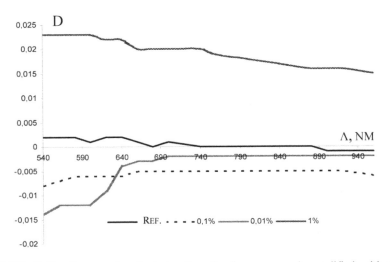

FIGURE 8.32 Curves of optical density of reference sample modified with Cu/C nanocomposites in concentration 1%, 0.1%, 0.01%.

Examination of curves of sample optical density demonstrated that when the NS concentration was 1% from polycarbonate mass, the visible-light spectrum was absorbed by about 4.2 per cent more when compared with the reference sample. When the nanocomposite concentration was 0.1 percent, the absorption decreased by 0.7 per cent. When the concentration was 0.01 per cent, the absorption decreased in the region 540–600 nm by 2.3 per cent, and in the region 640–960 nm – by 0.5 per cent.

During the microscopic investigation of the samples the schematic picture of the structures formed was obtained at 20-mcm magnification. The results are given in Figure 8.33. From the schematic pictures it is seen that volumetric structures of regular shape surrounded by micellae were formed in polycarbonate modified with 0.01% Cu/C nanocomposite. When 0.1% of nanocomposite was introduced, the linear structures distorted in space and surrounded by micellae were formed. When Cu/C nanocomposite concentration was 1%, large aggregates were not observed in polycarbonate.

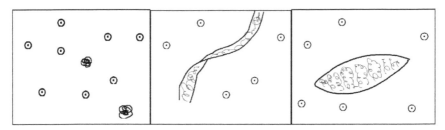

FIGURE 8.33   schematic picture of structure formation in polycarbonate modified with Cu/C nanocomposites in concentrations 1%, 0.1%, 0.01% (From left to the right), 20-mcm magnification.

Apparently, the decrease in nanocomposite concentration in polycarbonate can result in the formation of self-organizing structures of bigger size. In Ref. [3], there is an hypothesis on the transfer of nanocomposite oscillations onto the molecules of polymeric composition, and the intensity of bands in IR spectra which sharply increases even after the introduction of supersmall quantities of nanocomposites. In our case, this hypothesis was checked on modified and non-modified samples of polycarbonate films.

In Figure 8.34, the IR spectra of polycarbonate and polycarbonate modified with 0.001% of Cu/C nanocomposite can be observed.

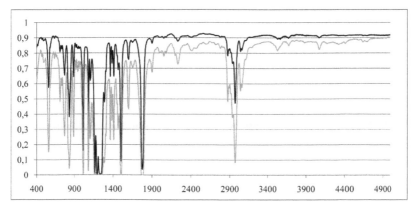

FIGURE 8.34   ir spectra of reference sample (upper) and polycarbonate modified with Cu/C nanocomposites in concentration 0.001% (Lower).

As seen in IR spectra, the intensity increases in practically all regions, indicating the influence of oscillations of Cu/C nanocomposite on all the systems. The most vivid changes in the band intensity are observed at 557 cm^{-1}, in the region 760–930 cm^{-1}, at 1,400 cm^{-1}, 1,600 cm^{-1}, and 2,970 cm^{-1}. Thus, polycarbonate self-organization under the influence of Cu/C nanocomposite takes place with the participation of certain bonds, for which the intensity of absorption bands goes up. At the same time, the formation of new phases is possible which usually results in thermal capacity increase. However, thermal conductivity can decrease due to the formation of defective regions between the aggregates formed. Table 8.11 contains the results of thermal–physical characteristics studied.

TABLE 8.11   Thermal–physical characteristics of polycarbonate and its modified analogs

Parameters	Nanocomposite content in polycarbonate (%)			
	0%, ref.	1%	0.1%	0.01%
m' 10^{-3}, kg	1.9	1.955	1.982	1.913
h'10^{-3}, m	9.285	9.563	9.76	9.432
Csp, J/kg K	1440	1028	1400	1510
Λ	0.517	0.503	0.487	0.448

It is demonstrated that when the NS concentration in the material decreases, thermal capacity goes up which is confirmed by the results of previous investigations. Thermal conductivity decline, when the NS concentration decreases, is apparently caused by the material defectiveness When Cu/C nanocomposites are introduced into the modified material, the NS can be considered as the generator of molecules excitation, which results in wave process in the material. It is found that polycarbonate modification with metal/carbon containing nanocomposites results in the changes in polycarbonate structure influencing its optical and thermal–physical properties.

KEYWORDS

- Hydroxyfullerene molecules
- quantum-chemical investigation
- Semiempirical method

REFERENCES

1. Schmidt, M. W.; Baldridge, K. K.; Boatz, J. A.; et al., J.Comput. Chem., 1993, v.14, p. 1347–1363.
2. Copley, J. R. D.; Neumann, D. A.; Cappelletti, R. L.; Kamitakahara, W. A.; J. Phys. Chem. Solids. 1992. V. **53**. N 11. P. 1353–1371.
3. Sh. Lo, V. Li, *Russian chemical journal*, *5*, 40–48 (**1999**)
4. A.N. Ponomarev, *Constructional materials*, *6*, 69–71 (**2007**).
5. V.A. Krutikov, A.A. Didik, G.I. Yakovlev, V.I. Kodolov, A.Yu. Bondar, *Alternative power engineering and ecology*, *4*, 36–41 (**2005**)
6. A.A. Granovsky, PC GAMESS version 7.1.5 (http://classic.chem.msu.su/gran/gamess/index.html). M.W.Schmidt, K.K.Baldridge, J.A.Boatz et al. *J.Comput.Chem.* 14, 1347–1363 (**1993**)
7. Bürgi, T.; Droz, T.; Leutwyler, S.; *Chemical Physics Letters* **1995**, *246*, 291–299.
8. Matsumoto, Y.; Ebata, T.; Mikami, N. J.; *Chem. Phys.* **1998**, *109*, 6303–6311.
9. Schemmel, D.; Schütz, M. J.; *Chem. Phys.* **2008**, *129*, 034301.
10. Famulari, A.; Raimondi, M.; Sironi, M.; Gianinetti, E.; *Chemical Physics* **1998**, *232*, 275–287.
11. Hedberg, K.; Hedberg, L.; Bethune, D.S.; et al., *Science* **1991**, *254*, 410–412.
12. Copley, J. R. D.; Neumann, D. A.; Cappelletti, R. L.; Kamitakahara, W. A.; *J Phys Chem Solids* **1992**, *53*, 1353–1371.
13. Lipanov, A. M.; Kodolov, V. I.; Khokhriakov, N. V.; Didik, A. A.; Kodolova, V.V.; Semakina, N. V.; *Alternative power engineering and ecology*, **2005**, № 2. P. 58–63
14. M. Valiev, E.J. Bylaska, N. Govind, K. Kowalski, T.P. Straatsma, H. J. J. van Dam, D. Wang, J. Nieplocha, E. Apra, T.L. Windus, W.A. de Jong, Comput. *Phys. Commun.*, *181*, 9, 1477–1489 (**2010**)
A. Famulari, M. Raimondi, M. Sironi, E. Gianinetti *Chemical Physics*, *232*, 289–298 (**1998**)
15. Spectral database for organic compounds AIST. http://riodb01.ibase.aist.go.jp/sdbs/cgi-bin/cre_index.cgi?lang=eng
16. HyperChem 8 manual. Computational Chemistry, **2002**. p. 149, p. 369.
17. Seeger, D. M.; Korzeniewski, C.; Kowalchyk, W.; Jounal of Physical Chemistry, **1991**. Iss. 95. Pp. 68–71.
18. Kazicina, L. A.; Kupletskaya, N. B.; Application of UV-, IR- and NMR spectroscopy in organic chemistry.–M.: Chemistry, 1971–264p.

19. Nakamoto, K.; Infrared spectra of inorganic and coordination compounds. M.: 1966–412p.
20. Kodolov, V. I.; Khokhriakov, N. V.; Chemical physics of formation and transformation processes of nanostrcutures and nanosystems. –Izhevsk: Izhevsk State Agricultural Academy, **2009**, V.*1* (*365*p), V.*2* (*4153*).
21. Kodolov, V.I.; Khokhriakov, N. V.; Trineeva, V. V.; Blagodatskikh I.I. Activity of nanostructures and its expression in nanoreactors of polymeric matrixes and active media. Chemical physics and mesoscopy, **2008**. V.**10**. -Iss. 4. Pp. 448–460.
Panfilov, B. F.; Composite materials: production, application, market tendencies. Polymeric materials, **2010**.-Iss. 2–3. Pp. 40–43.
22. Bulgakov, V. K.; Kodolov, V. I.; Lipanov, A. M.; Modeling of polymeric materials combustion. –M.: Chemistry, **1990**. –238p.
23. Shuklin, S.G.; Kodolov, V. I.; Larionov, K. I.; Tyurin, S. A.; Physical and chemical processes in modified two-layer fire and heat-resistant epoxy-polymers under the action of fire sources Physics of combustion and explosion. **1995**. V. **31**.-Iss. 2.-P. 73–79.
24. Surface functionalization of multi-wall carbon nanotubes N.V. Glebova, A.A. Nechitalov Journal "Journal of Technical Physics", **2010**, V. **36**, iss. 19, p. 12–15.
25. Patent 2393110 Russia Technique of obtaining carbon metal containing nanostructures Kodolov, V. I.; Vasilchenko, Yu. M.; Akhmetshina, L. F.; Shklyaeva, D. A.; Trineeva, V. V.; Sharipova, A. G.; Volkova, E. G.; Ulyanov, A. L.; Kovyazina, O. A.; declared on 17.10.**2008**, published on 27.06.**2010**.
26. Patent 2337062 Russia Technique of obtaining carbon nanostructures from organic compounds and metal containing substances Kodolov, V. I.; Kodolova, V. V.; (Trineeva), Semakina, N. V.; Yakovlev, G. I.; Volkova, E. G.; et al.; declared on 28.08.**2006**, published on 27.10.**2008**.
27. Kodolov, V. I.; Trineeva, V. V.; Kovyazina, O. A.; Vasilchenko, Yu. M.; Production and application of metal carbon nanocomposites. In book "The problems of nanochemistry for the creation of new materials" Torun. Poland: IEPMD, **2012**. Pp. 17–22.
28. V. I. Kodolov, V. V. Trineeva. Perspectives of idea development about nanosystems self-organization in polymeric matrixes. In book "The problems of nanochemistry for the creation of new materials" Torun. Poland: IEPMD, **2012**. Pp. 75–100.

CHAPTER 9

CALCULATING THE INTERNAL STRUCTURE AND THE EQUILIBRIUM CONFIGURATION (SHAPE) OF SEPARATE NONINTERACTING NANOPARTICLES BY THE MOLECULAR MECHANICS AND DYNAMICS INTERACTIONS OF NANOSTRUCTURAL ELEMENTS

A. V. VAKHRUSHEV and A. M. LIPANOV

CONTENTS

9.1   INTRODUCTION

The properties of a nanocomposite are determined by the structure and properties of the nanoelements, which form it. One of the main tasks in making nanocomposites is building the dependence of the structure and shape of the nanoelements forming the basis of the composite on their sizes. This is because with an increase or a decrease in the specific size of nanoelements (nanofibers, nanotubes, nanoparticles, and so on), their physical—mechanical properties such as coefficient of elasticity, strength, deformation parameter, and so on, are varying over one order [1–5].

The calculations and experiments show that this is primarily due to a significant rearrangement (which is not necessarily monotonous) of the atomic structure and the shape of the nanoelement. The experimental investigation of the aforementioned parameters of the nanoelements is technically complicated and laborious because of their small sizes. In addition, the experimental results are often inconsistent. In particular, some authors have pointed to an increase in the distance between the atoms adjacent to the surface in contrast to the atoms inside the nanoelement, whereas others observe a decrease in the aforementioned distance [6].

Thus, further detailed systematic investigations of the problem with the use of theoretical methods, i.e., mathematical modeling, are required. The atomic structure and the shape of nanoelements depend both on their sizes and on the methods of obtaining which can be divided into two main groups:

1. Obtaining nanoelements in the atomic coalescence process by "assembling" the atoms and by stopping the process when the nanoparticles grow to a desired size (the so-called "bottom—up" processes). The process of the particle growth is stopped by the change of physical or chemical conditions of the particle formation, by cutting off supplies of the substances that are necessary to form particles, or because of the limitations of the space where nanoelements form.
2. Obtaining nanoelements by breaking or destructing more massive (coarse) formations to the fragments of the desired size (the so-called "up—down" processes).

In fact, there are many publications describing the modeling of the "bottom-up" processes [7–8], whereas the "up-down" processes have been studied very little. Therefore, the objective of this work is the investigation of the regularities of the changes in the structure and shape of nanoparticles formed in the destruction ("up-down") processes depending on the nanoparticle sizes, and building up theoretical dependences describing the aforementioned parameters of nanoparticles.

When the characteristics of powder nanocomposites are calculated it is also very important to take into account the interaction of the nanoelements since the changes in their original shapes and sizes in the interaction process during the formation (or usage) of the nanocomposite can lead to a significant change in its properties and a cardinal structural rearrangement. In addition, the experimental investigations show the appearance of the processes of ordering and self-assembling leading to a more organized form of a nanosystem [9–15].

In general, three main processes can be distinguished: the first process is because of the regular structure formation at the interaction of the nanostructural elements with the surface where they are situated; the second one arises from the interaction of the nanostructural elements with one another; the third process takes place because of the influence of the ambient medium surrounding the nanostructural elements. The ambient medium influence can have "isotropic distribution" in the space or it can be presented by the action of separate active molecules connecting nanoelements to one another in a certain order. The external action significantly changes the original shape of the structures formed by the nanoelements. For example, the application of the external tensile stress leads to the "stretch" of the nanoelement system in the direction of the maximal tensile stress action; the rise in temperature, vice versa, promotes a decrease in the spatial anisotropy of the nanostructures [10]. Note that in the self-organizing process, parallel with the linear moving, the nanoelements are in rotary movement. The latter can be explained by the action of moment of forces caused by the asymmetry of the interaction force fields of the nanoelements, by the presence of the "attraction" and "repulsion" local regions on the nanoelement surface, and by the "non-isotropic" action of the ambient as well.

The aforementioned phenomena play an important role in nanotechnological processes. They allow developing nanotechnologies for the formation of nanostructures by the self-assembling method (which is based on self-organizing processes) and building up complex spatial nanostructures consisting of different nanoelements (nanoparticles, nanotubes, fullerenes, super-molecules, and so on) [15]. However, in a number of cases, the tendency towards self-organization interferes with the formation of a desired nanostructure. Thus, the nanostructure arising from the self-organizing process is, as a rule, "rigid" and stable against external actions. For example, the "adhesion" of nanoparticles interferes with the use of separate nanoparticles in various nanotechnological processes, the uniform mixing of the nanoparticles from different materials and the formation of nanocomposite with desired properties. In connection with this, it is important to model the processes of static and dynamic interaction of the nanostructure elements. In this case, it is essential to take into consideration the interaction force moments of the nanostructure elements, which causes the mutual rotation of the nanoelements.

The investigation of the aforementioned dependences based on the mathematical modeling methods requires the solution of the aforementioned problem on the atomic level. This requires large computational aids and computational time, which makes the development of economical calculation methods urgent. The objective of this work was the development of such a technique. This chapter gives results of the studies of problems of numeric modeling within the framework of molecular mechanics and dynamics for investigating the regularities of the amorphous phase formation and the nucleation and spread of the crystalline or hypocrystalline phases over the entire nanoparticle volume depending on the process parameters, nanoparticles sizes, and thermodynamic conditions of the ambient. In addition, the method for calculating the interactions of nanostructural elements is offered, which is based on the potential built up with the help of the approximation of the numerical calculation results using the method of molecular dynamics of the pairwise static interaction of nanoparticles. Based on the potential of the pairwise interaction of the nanostructure elements, which takes into account forces and moments of forces, the method for calculating the ordering and self-organizing processes has been developed.

The investigation results on the self-organization of the system consisting of the presence of two or more particles and the analysis of the equilibrium stability of various types of nanostructures has been carried out. These results are a generalization of the authors' research in [16–24]. A more detailed description of the problem can be obtained in these works.

9.2 PROBLEM STATEMENT

The problem on calculating the internal structure and the equilibrium configuration (shape) of separate non-interacting nanoparticles by the molecular mechanics and dynamics methods has two main stages:

1. The "initiation" of the task, i.e., the determination of the conditions under which the process of the nanoparticle shape and structure formation begins.
2. The process of the nanoparticle formation.

Note that the original coordinates and initial velocities of the nanoparticle atoms should be determined from the calculation of the macroscopic parameters of the destructive processes at static and dynamic loadings taking place on both the nanoscale and the macroscale. Therefore, in the general case, the coordinates and velocities are the result of solving the problem of modeling physical—mechanical destruction processes at different structural levels. This problem due to its enormity and complexity is not considered in this chapter. The detailed description of its statement and the numerical results of its solution are given in the works of the authors [16–19].

The problem of calculating the interaction of ordering and self-organization of the nanostructure elements includes three main stages: the first stage is building the internal structure and the equilibrium configuration (shape) of each separate non-interacting nanostructure element; the second stage is calculating the pairwise interaction of two nanostructure elements; and the third stage is establishing the regularities of the spatial structure and evolution with time of the nanostructure as a whole. Let us consider the aforementioned problems in sequence.

9.2.1 THE CALCULATION OF THE INTERNAL STRUCTURE AND THE SHAPE OF THE NON-INTERACTING NANOELEMENT

The initialization of the problem is in giving the initial coordinates and velocities of the nanoparticle atoms.

$$\vec{x}_i = \vec{x}_{i0}, \vec{V}_i = \vec{V}_{i0}, \ t = 0, \vec{x}_i \subset \Omega_k , \tag{9.1}$$

where $\vec{\mathbf{x}}_{i0}, \vec{\mathbf{x}}_i$ are original and current coordinates of the i th atom; $\vec{\mathbf{V}}_{i0}, \vec{\mathbf{V}}_i$ are initial and current velocities of the i th atom, respectively; Ω_k is an area occupied by the nanoelement.

The problem of calculating the structure and the equilibrium configuration of the nanoelement will be carried out with the use of the molecular dynamics method taking into consideration the interaction of all the atoms forming the nanoelement. Since, at the first stage of the solution, the nanoelement is not exposed to the action of external forces, it is taking the equilibrium configuration with time, which is further used for the next stage of calculations.

At the first stage, the movement of the atoms forming the nanoparticle is determined by the set of Langevin differential equations at the boundary conditions Eq. (9.1) [25]

$$m_i \cdot \frac{d\vec{\mathbf{V}}_i}{dt} = \sum_{j=1}^{N_k} \vec{\mathbf{F}}_{ij} + \vec{\mathbf{F}}_i(t) - \alpha_i m_i \vec{\mathbf{V}}_i, \qquad i = 1, 2, \dots, N_k,$$

$$\frac{d\vec{\mathbf{x}}_i}{dt} = \vec{\mathbf{V}}_i,$$

(9.2)

where N_k is the number of atoms forming each nanoparticle, m_i is the mass of the i th atom, α_i is the "friction" coefficient in the atomic structure, and $\vec{\mathbf{F}}_i(t)$ is a random set of forces at a given temperature which is given by Gaussian distribution.

Usually, the interatomic interaction forces are potential and determined by the relation

$$\vec{\mathbf{F}}_{ij} = -\sum_{1}^{n} \frac{\partial \Phi(\vec{\rho}_{ij})}{\partial \vec{\rho}_{ij}}, i = 1, 2, \dots, N_k, \ j = 1, 2, \dots, N_k,$$

(9.3)

where $\vec{\rho}_{ij}$ is a radius-vector determining the position of the i th atom relative to the j th atom; $\Phi(\vec{\rho}_{ij})$ is a potential depending on the mutual positions of all the atoms; n is the number of interatomic interaction types.

In the general case, the potential $\Phi(\vec{\rho}_{ij})$ is given in the form of the sum of several components corresponding to different interaction types:

$$\Phi(\vec{\rho}_{ij}) = \Phi_{cb} + \Phi_{va} + \Phi_{ta} + \Phi_{pg} + \Phi_{vv} + \Phi_{es} + \Phi_{hb}.$$

(9.4)

Here the following potentials are implied: Φ_{cb}—of chemical bonds, Φ_{va}—of valence angles, Φ_{ta}—of torsion angles; Φ_{pg}—of flat groups; Φ_{vv}—of Van der Waals contacts; Φ_{es}—of electrostatics; and Φ_{hb}—of hydrogen bonds.

The aforementioned addends have different functional forms. The parameter values for the interaction potentials are determined based on the experiments (crystallography, spectral, calorimetric, and so on) and quantum calculations [25].

Giving original coordinates (and forces of atomic interactions) and velocities of all the atoms of each nanoparticle in accordance with Eq. (9.2), at the start time, we find the change of the coordinates and the velocities of each nanoparticle atoms with time from the equation of motion Eq. (9.1). Since the nanoparticles are not exposed to the action of external forces, they take some atomic equilibrium configuration with time that we will use for the next calculation stage.

9.3   THE CALCULATION OF THE PAIRWISE INTERACTION OF THE TWO NANOSTRUCTURE ELEMENTS

At this stage of solving the problem, we consider two interacting nanoelements. First, let us consider the problem statement for symmetric nanoelements, and then for arbitrary shaped nanoelements.

First of all, let us consider two symmetric nanoelements situated at the distance S from one another (Figure 9.1) at the initial conditions

$$\vec{x}_i = \vec{x}_{i0}, \vec{V}_i = 0, \; t = 0, \; \vec{x}_i \subset \Omega_1 \bigcup \Omega_2 \,, \qquad (9.5)$$

where Ω_1, Ω_2 are the areas occupied by the first and the second nanoparticle, respectively.

We obtain the coordinates $\vec{x}_{i0}$ from Eq. (9.2) solution at initial conditions (9.1). It allows calculating the combined interaction forces of the nanoelements

$$\vec{F}_{b1} = -\vec{F}_{b2} = \sum_{i=1}^{N_1} \sum_{j=1}^{N_2} \vec{F}_{ij} \,, \qquad (9.6)$$

where i, j are the atoms and N_1, N_2 are the numbers of atoms in the first and the second nanoparticles, respectively.

Forces $\vec{F}_{ij}$ are defined from Eq. (9.3).

In the general case, the force magnitude of the nanoparticle interaction $\left|\vec{F}_{bi}\right|$ can be written as product of functions depending on the sizes of the nanoelements and the distance between them:

$$\left|\vec{F}_{bi}\right| = \Phi_{11}(S_c) \times \Phi_{12}(D) \tag{9.7}$$

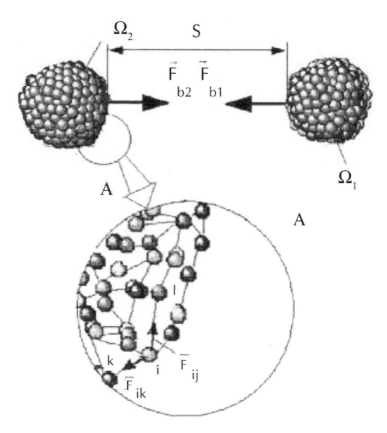

FIGURE 9.1 Scheme of the nanoparticle interaction; A—an enlarged view of the nanoparticle fragment.

The $\vec{F}_{bi}$ vector direction is determined by the direction cosines of a vector connecting the centers of the nanoelements.

Now, let us consider two interacting asymmetric nanoelements situated at the distance S_c between their centers of mass (Figure 9.2) and oriented at certain specified angles relative to each other.

In contrast to the earlier problem, the interatomic interaction of the nanoelements leads not only to the relative displacement of the nanoelements but also to their rotation as well. Consequently, in the general case, the sum of all the forces of the interatomic interactions of the nanoelements is brought to the principal vector of forces $\vec{F}_c$ and the principal moment $\vec{M}_n$

$$\vec{F}_c = \vec{F}_{b1} = -\vec{F}_{b2} = \sum_{i=1}^{N_1} \sum_{j=1}^{N_2} \vec{F}_{ij} \, , \tag{9.8}$$

$$\vec{M}_c = \vec{M}_{c1} = -\vec{M}_{c2} = \sum_{i=1}^{N_1} \sum_{j=1}^{N_2} \vec{\rho}_{cj} \times \vec{F}_{ij} \, , \tag{9.9}$$

where $\vec{\rho}_{cj}$ is a vector connecting points c and j.

The main objective of this calculation stage is building the dependences of the forces and moments of the nanostructure—nanoelement interactions on the distance S_c between the centers of mass of the nanostructure nanoelements, on the angles of mutual orientation of the nanoelements $\Theta_1, \Theta_2, \Theta_3$ (shapes of the nanoelements) and on the characteristic size D of the nanoelement. In the general case, these dependences can be given in the form

$$\vec{F}_{bi} = \vec{\Phi}_F (S_c, \Theta_1, \Theta_2, \Theta_3, D) \, , \tag{9.10}$$

$$\vec{M}_{bi} = \vec{\Phi}_M (S_c, \Theta_1, \Theta_2, \Theta_3, D) \, , \tag{9.11}$$

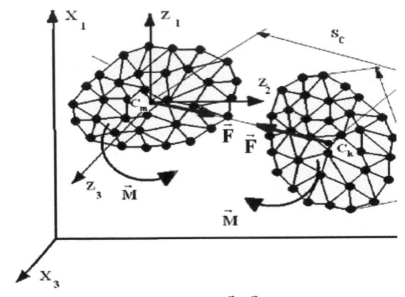

FIGURE 9.2 Two interacting nanoelements; $\vec{M}$, $\vec{F}$ are the principal moment and the principal vector of the forces, respectively.

For spherical nanoelements, the angles of the mutual orientation do not influence the force of their interaction; therefore, in Eq. (9.12), the moment is zero.

In the general case, functions in Eqs. (9.11) and (9.12) can be approximated by analogy with Eq. (9.8) as the product of functions $S_0, \Theta_1, \Theta_2, \Theta_3, D$, respectively. For the further numerical solution of the problem of the self-organization of nanoelements, it is sufficient to give the aforementioned functions in their tabular form and to use the linear (or non-linear) interpolation of them in space.

9.4   PROBLEM FORMULATION FOR INTERACTION OF SEVERAL NANOELEMENTS

When the evolution of the nanosystem as whole (including the processes of ordering and self-organization of the nanostructure nanoelements) is investigated, the movement of each system nanoelement is considered as the movement of a single whole. In this case, the translational motion of the center of mass of each nanoelement is given in the coordinate system X_1, X_2, X_3, and the nanoelement rotation is described in the coordinate system Z_1, Z_2, Z_3, which is related to the center of mass of the nanoelement (Figure 9.2). The system of equations describing the aforementioned processes has the form.

$$
\begin{cases}
M_k \dfrac{d^2 X_1^k}{dt^2} = \displaystyle\sum_{j=1}^{N_e} F_{X_1}^{kj} + F_{X_1}^{ke}, \\[2mm]
M_k \dfrac{d^2 X_2^k}{dt^2} = \displaystyle\sum_{j=1}^{N_e} F_{X_2}^{kj} + F_{X_2}^{ke}, \\[2mm]
M_k \dfrac{d^2 X_3^k}{dt^2} = \displaystyle\sum_{j=1}^{N_e} F_{X_3}^{kj} + F_{X_3}^{ke}, \\[2mm]
J_{Z_1}^k \dfrac{d^2 \Theta_1^k}{dt^2} + \dfrac{d\Theta_2^k}{dt} \cdot \dfrac{d\Theta_3^k}{dt}(J_{Z_3}^k - J_{Z_2}^k) = \displaystyle\sum_{j=1}^{N_e} M_{Z_1}^{kj} + M_{Z_1}^{ke}, \\[2mm]
J_{Z_2}^k \dfrac{d^2 \Theta_2^k}{dt^2} + \dfrac{d\Theta_1^k}{dt} \cdot \dfrac{d\Theta_3^k}{dt}(J_{Z_1}^k - J_{Z_3}^k) = \displaystyle\sum_{j=1}^{N_e} M_{Z_2}^{kj} + M_{Z_2}^{ke}, \\[2mm]
J_{Z_3}^k \dfrac{d^2 \Theta_3^k}{dt^2} + \dfrac{d\Theta_2^k}{dt} \cdot \dfrac{d\Theta_1^k}{dt}(J_{Z_2}^k - J_{Z_1}^k) = \displaystyle\sum_{j=1}^{N_e} M_{Z_3}^{kj} + M_{Z_3}^{ke},
\end{cases}
\tag{9.12}
$$

where x_i^k, Θ_i^k are coordinates of the centres of mass and angles of the spatial orientation of the principal axes Z_1, Z_2, Z_3 of nanoelements; $F_{X_1}^{kj}, F_{X_2}^{kj}, F_{X_3}^{kj}$ are the interaction forces of nanoelements; $F_{X_1}^{ke}, F_{X_2}^{ke}, F_{X_3}^{ke}$ are external forces acting on nanoelements; N_e is the number of nanoelements; M_k is a mass of a nanoelement; $M_{Z_1}^{ke}, M_{Z_2}^{ke}, M_{Z_3}^{ke}$ are the moment of forces of the nanoelement interaction; $M_{Z_1}^{ke}, M_{Z_2}^{ke}, M_{Z_3}^{ke}$ are the external moments acting on nanoelements; and $J_{Z_1}^k, J_{Z_2}^k, J_{Z_3}^k$ are moments of inertia of a nanoelement.

The initial conditions for the system of Eqs. (9.13) and (9.14) have the form

$$
\vec{X}^k = \vec{X}_0^k; \ \ \Theta^k = \Theta_0^k; \ \vec{V}^k = \vec{V}_0^k; \ \frac{d\Theta^k}{dt} = \frac{d\Theta_0^k}{dt}; \ t = 0,
\tag{9.13}
$$

9.5   NUMERICAL PROCEDURES AND SIMULATION TECHNIQUES

In the general case, the problem formulated in the previous sections has no analytical solution at each stage; therefore, numerical methods for solving are used, as a rule. In this chapter, for the first stages, the numerical integration of the equation of motion of the nanoparticle atoms in the relaxation process are used in accordance with Verlet scheme [26]:

$$\vec{x}_i^{n+1} = \vec{x}_i^n + \Delta t\ \vec{V}_i^n + \left((\Delta t)^2 / 2m_i\right)\left(\sum_{j=1}^{N_k}\vec{F}_{ij} + \vec{F}_i - \alpha_i m_i \vec{V}_i^n\right)^n, \tag{9.14}$$

$$\vec{V}_i^{n+1} = (1 - \Delta t\alpha_i)\vec{V}_i^n + (\Delta t / 2m_i)((\sum_{j=1}^{N_k}\vec{F}_{ij} + \vec{F}_i)^n + (\sum_{j=1}^{N_k}\vec{F}_{ij} + \vec{F}_i)^{n+1}), \tag{9.15}$$

where $\vec{x}_i^n$, $\vec{V}_i^n$ are a coordinate and a velocity of the i th atom at the n th step with respect to the time, and Δt is a step with respect to the time.

The solution of Eq. (9.13) also requires the application of numerical methods of integration. In the present work, Runge–Kutta method [27] is used for solving Eq. (9.13).

$$(X_i^k)_{n+1} = (X_i^k)_n + (V_i^k)_n\Delta t + \frac{1}{6}(\mu_{1i}^k + \mu_{2i}^k + \mu_{3i}^k)\Delta t, \tag{9.16}$$

$$(V_i^k)_{n+1} = (V_i^k)_n + \frac{1}{6}(\mu_{1i}^k + 2\mu_{2i}^k + 2\mu_{3i}^k + \mu_{4i}^k). \tag{9.17}$$

$$\mu_{1i}^k = \Phi_i^k\left(t_n; \left(X_i^k\right), \dots; \left(V_i^k\right) \dots\right)\Delta t,$$

$$\mu_{2i}^k = \Phi_i^k(t_n + \frac{\Delta t}{2}; (X_i^k + V_i^k\frac{\Delta t}{2})_n, \dots; (V_i^k)_n + \frac{\mu_{1i}^k}{2}, \dots)\Delta t,$$

$$\mu_{3i}^k = \Phi_i^k(t_n + \frac{\Delta t}{2}; (X_i^k + V_i^k\frac{\Delta t}{2} + \mu_{1i}^k\frac{\Delta t}{4})_n, \dots; (V_i^k)_n + \frac{\mu_{2i}^k}{2}, \dots)\Delta t,$$

$$\mu_{4i}^k = \Phi_i^k(t_n + \Delta t; (X_i^k + V_i^k \Delta t + \mu_{2i}^k \frac{\Delta t}{2})_n, ...; (V_i^k)_n + \mu_{2i}^k, ...)\Delta t .$$

$$\mu_{4i}^k = \Phi_i^k(t_n + \Delta t; (X_i^k + V_i^k \Delta t + \mu_{2i}^k \frac{\Delta t}{2})_n, ...; (V_i^k)_n + \mu_{2i}^k, ...)\Delta t \qquad (9.18)$$

$$\Phi_i^k = \frac{1}{M_k}(\sum_{j=1}^{N_e} F_{X_3}^{kj} + F_{X_3}^{ke}) \qquad (9.19)$$

$$(\Theta_i^k)_{n+1} = (\Theta_i^k)_n + (\frac{d\Theta_i^k}{dt})_n \Delta t + \frac{1}{6}(\lambda_{1i}^k + \lambda_{2i}^k + \lambda_{3i}^k)\Delta t \qquad (9.20)$$

$$(\frac{d\Theta_i^k}{dt})_{n+1} = (\frac{d\Theta_i^k}{dt})_n + \frac{1}{6}(\lambda_{1i}^k + 2\lambda_{2i}^k + 2\lambda_{3i}^k + \lambda_{4i}^k) \qquad (9.21)$$

$$\lambda_{1i}^k = \Psi_i^k(t_n; (\Theta_i^k)_n, ...; (\frac{d\Theta_i^k}{dt})_n, ...)\Delta t ,$$

$$\lambda_{2i}^k = \Psi_i^k(t_n + \frac{\Delta t}{2}; (\Theta_i^k + \frac{d\Theta_i^k}{dt}\frac{\Delta t}{2})_n, ...; (\frac{d\Theta_i^k}{dt})_n + \frac{\lambda_{1i}^k}{2}, ...)\Delta t ,$$

$$\lambda_{3i}^k = \Psi_i^k(t_n + \frac{\Delta t}{2}; (\Theta_i^k + \frac{d\Theta_i^k}{dt}\frac{\Delta t}{2} + \lambda_{1i}^k \frac{\Delta t}{4})_n, ...; (\frac{d\Theta_i^k}{dt})_n + \frac{\lambda_{2i}^k}{2}, ...)\Delta t , \qquad (9.22)$$

$$\lambda_{4i}^k = \Psi_i^k(t_n + \Delta t; (\Theta_i^k + \frac{d\Theta_i^k}{dt}\Delta t + \lambda_{2i}^k \frac{\Delta t}{2})_n, ...; (\frac{d\Theta_i^k}{dt})_n + \lambda_{2i}^k, ...)\Delta t .$$

$$\Psi_1^k = \frac{1}{J_{Z_1}^k}(-\frac{d\Theta_2^k}{dt} \cdot \frac{d\Theta_3^k}{dt}(J_{Z_3}^k - J_{Z_2}^k) + \sum_{j=1}^{N_e} M_{Z_1}^{kj} + M_{Z_1}^{ke}),$$

$$\Psi_2^k = \frac{1}{J_{Z_2}^k}(-\frac{d\Theta_1^k}{dt} \cdot \frac{d\Theta_3^k}{dt}(J_{Z_1}^k - J_{Z_3}^k) + \sum_{j=1}^{N_e} M_{Z_2}^{kj} + M_{Z_2}^{ke}), \qquad (9.23a)$$

$$\Psi_3^k = \frac{1}{J_{Z_3}^k}(-\frac{d\Theta_1^k}{dt} \cdot \frac{d\Theta_2^k}{dt}(J_{Z_1}^k - J_{Z_2}^k) + \sum_{j=1}^{N_e} M_{Z_3}^{kj} + M_{Z_3}^{ke}),$$

where $i=1, 2, 3; \; k=1, 2, \; N_e$

9.6   RESULTS AND DISCUSSIONS

Let us consider the realization of the aforementioned procedure taking the calculation of the metal nanoparticle as an example.

The potentials of the atomic interaction of Morse (Eq. (9.23b)) and Lennard-Johns (Eq. (9.24)) were used in the following calculations

$$\Phi(\vec{\rho}_{ij})_m = D_m \left(\exp(-2\lambda_m(|\vec{\rho}_{ij}| - \rho_0)) - 2\exp(-\lambda_m(|\vec{\rho}_{ij}| - \rho_0)) \right), \qquad (9.23b)$$

$$\Phi(\vec{\rho}_{ij})_{LD} = 4\varepsilon \left[\left(\frac{\sigma}{|\vec{\rho}_{ij}|} \right)^{12} - \left(\frac{\sigma}{|\vec{\rho}_{ij}|} \right)^{6} \right], \qquad (9.24)$$

where $D_m, \lambda_m, \rho_0, \varepsilon, \sigma$ are the constants of the materials studied.

For sequential and parallel solving the molecular dynamics equations, the program package developed at Applied Mechanics Institute, the Ural Branch of the Russian Academy of Sciences, and the advanced program package NAMD developed at the University of Illinois, and Beckman Institute (USA) by the Theoretical Biophysics Group were used. The graphic imaging of the nanoparticle calculation results was carried out with the use of the program package VMD.

9.6.1   STRUCTURE AND FORMS OF NANOPARTICLES

At the first stage of the problem, the coordinates of the atoms positioned at the ordinary material lattice points (Figure 9.3, (1)) were taken as the original coordinates. During the relaxation process, the initial atomic system is rearranged into a new "equilibrium" configuration (Figure 9.3, (2)) in accordance with the calculations based on Eq. (9.6)–(9.9), which satisfies the condition when the system potential energy is approaching the minimum (Figure 9.3, the plot).

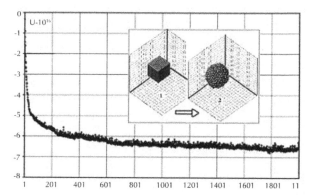

FIGURE 9.3 The initial crystalline (1) and cluster (2) structures of the nanoparticle consisting of 1331 atoms after relaxation; the plot of the potential energy U [J] variations for this atomic system in the relaxation process (n number of iterations with respect to the time).

After the relaxation, the nanoparticles can have quite diverse shapes: globe-like, spherical centered, spherical eccentric, spherical icosahedral nanoparticles, and asymmetric nanoparticles (Figure 9.4).

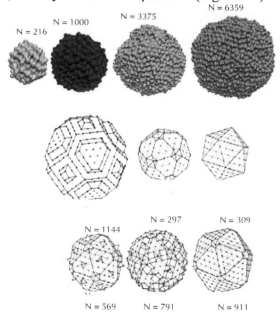

FIGURE 9.4 nanoparticles of diverse shapes, depending on the number of atoms they consist of.

In this case, the number of atoms N significantly determines the shape of a nanoparticle. Note, that symmetric nanoparticles are formed only at a certain number of atoms. As a rule, in the general case, the nanoparticle deviates from the symmetric shape in the form of irregular raised portions on the surface. Besides, there are several different equilibrium shapes for the same number of atoms. The plot of the nanoparticle potential energy change in the relaxation process (Figure 9.5) illustrates it.

As it follows from this figure, the curve has two areas: the area of the decrease of the potential energy and the area of its stabilization promoting the formation of the first nanoparticle equilibrium shape (1). Then, a repeated decrease in the nanoparticle potential energy and the stabilization area corresponding to the formation of the second nanoparticle equilibrium shape are observed (2). Between them, there is a region of the transition from the first shape to the second one (P). The second equilibrium shape is more stable due to the lesser nanoparticle potential energy. However, the first equilibrium shape also "exists" rather long in the calculation process. The change of the equilibrium shapes is especially characteristic of the nanoparticles with an "irregular" shape. The internal structure of the nanoparticles is of importance since their atomic structure significantly differs from the crystalline structure of the bulk materials: the distance between the atoms and the angles change, and the surface formations of different types appear. In Figure 9.6, the change of the structure of a two-dimensional nanoparticle in the relaxation process is shown.

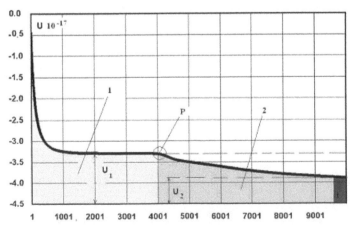

FIGURE 9.5 The plot of the potential energy change of the nanoparticle in the relaxation process.

1—a region of the stabilization of the first nanoparticle equilibrium shape; 2—a region of the stabilization of the second nanoparticle equilibrium shape; and P—a region of the transition of the first nanoparticle equilibrium shape into the second one.

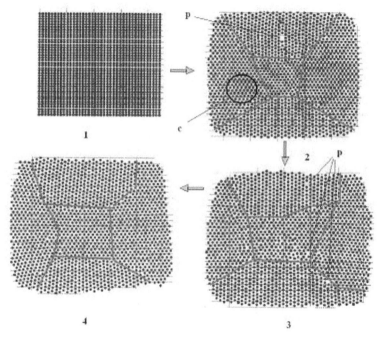

FIGURE 9.6 The change of the structure of a two-dimensional nanoparticle in the relaxation process: 1—the initial crystalline structure; 2, 3, 4—the nanoparticles structures which change in the relaxation process; p—pores; and c—the region of compression.

Figure 9.6 shows how the initial nanoparticle crystalline structure (position 1) is successively rearranging with time in the relaxation process (positions 2, 3, 4). Note that the resultant shape of the nanoparticle is not round, i.e., it has "remembered" the initial atomic structure. It is also of interest that in the relaxation process, in the nanoparticle, the defects in the form of pores (designation "p" in the figure) and the density fluctuation regions (designation "c" in the figure) have been formed, which are absent in the final structure.

9.7   NANOPARTICLES INTERACTION

Let us consider some examples of nanoparticles interaction. Figure 9.7 shows the calculation results demonstrating the influence of the sizes of the nanoparticles on their interaction force. One can see from the plot that the larger nanoparticles are attracted stronger, i.e., the maximal interaction force increases with the size growth of the particle. Let us divide the interaction force of the nanoparticles by its maximal value for each nanoparticle size, respectively. The obtained plot of the "relative" (dimensionless) force (Figure 9.8) shows that the value does not practically depend on the nanoparticle size as all the curves come close and can be approximated to one line.

Figure 9.9 displays the dependence of the maximal attraction force between the nanoparticles on their diameter that is characterized by nonlinearity and a general tendency towards the growth of the maximal force with the nanoparticle size growth.

The total force of the interaction between the nanoparticles is determined by multiplying of the two plots (Figures 9.8 and 9.9).

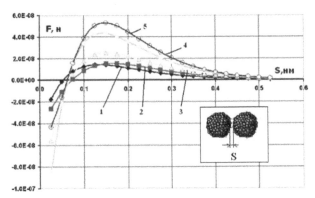

FIGURE 9.7   The dependence of the interaction force F [N] of the nanoparticles on the distance s [nm] between them and on the nanoparticle size: 1—d = 2.04; 2—d = 2.40; 3—d = 3.05; 4—d = 3.69; 5—d = 4.09 [nm].

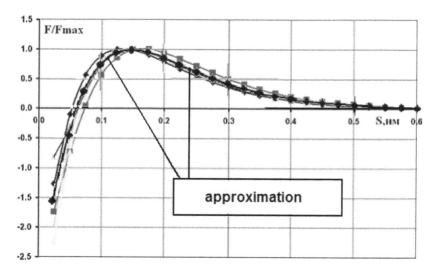

FIGURE 9.8 The dependence of the "relative" force $\overline{F}$ of the interaction of the nanoparticles on the distance S [nm] between them.

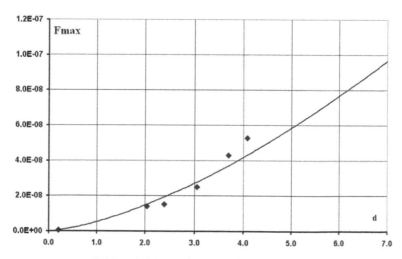

FIGURE 9.9 The dependence of the maximal attraction force F_{max} [N] the nanoparticles on the nanoparticle diameter d [nm].

Using the polynomial approximation of the curve in Figure 9.5 and the power mode approximation of the curve in Figure 9.6, we obtain

$$\overline{F} = (-1.13S^6 + 3.08S^5 - 3.41S^4 - 0.58S^3 + 0.82S - 0.00335)10^3 , \qquad (9.25)$$

$$F_{max} \cdot = 0.5 \cdot 10^{-9} \cdot d^{1.499} , \qquad (9.26)$$

$$F = F_{max} \cdot \overline{F} , \qquad (9.27)$$

where d and S are the diameter of the nanoparticles and the distance between them [nm], respectively; F_{max} is the maximal force of the interaction of the nanoparticles [N].

Dependences Eqs. (9.25)–(9.27) were used for the calculation of the nanocomposite ultimate strength for different patterns of nanoparticles' "packing" in the composite (Figure 9.10).

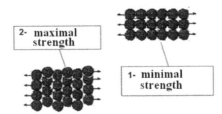

FIGURE 9.10   Different types of the nanoparticles' "packing" in the composite.

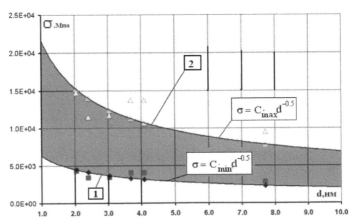

FIGURE 9.11   The dependence of the ultimate strength σ [MPa] of the nanocomposite formed by monodisperse nanoparticles on the nanoparticle sizes d [nm].

Figure 9.11 shows the dependence of the ultimate strength of the nano-composite formed by monodisperse nanoparticles on the nanoparticle sizes. One can see that with the decrease of the nanoparticle sizes, the ultimate strength of the nanomaterial increases, and vice versa. The calculations have shown that the nanocomposite strength properties are significantly influenced by the nanoparticles' "packing" type in the material. The material strength grows when the packing density of nanoparticles increases. It should be specially noted that the material strength changes in inverse proportion to the nanoparticle diameter in the degree of 0.5, which agrees with the experimentally established law of strength change of nanomaterials (the law by Hall-Petch) [18].

$$\sigma = C \cdot d^{-0.5},\qquad (9.28)$$

where $C = C_{max} = 2.17 \times 10^4$ is for the maximal packing density; $C = C_{min} = 6.4 \times 10^3$ is for the minimal packing density.

The electrostatic forces can strongly change force of interaction of nanoparticles. For example, numerical simulation of charged sodium (NaCl) nanoparticles system (Figure 9.12) has been carried out. Considered ensemble consists of eight separate nanoparticles. The nanoparticles interact due to Van-der-Waals and electrostatic forces.

Results of particles center of masses motion are introduced in Figure 9.13 representing trajectories of all nanoparticles included into system. It shows the dependence of the modulus of displacement vector $|R|$ on time. One can see that nanoparticle moves intensively at first stage of calculation process. At the end of numerical calculation, all particles have got new stable locations, and the graphs of the radius vector $|R|$ become stationary. However, the nanoparticles continue to "vibrate" even at the final stage of numerical calculations. Nevertheless, despite of "vibration", the system of nanoparticles occupies steady position.

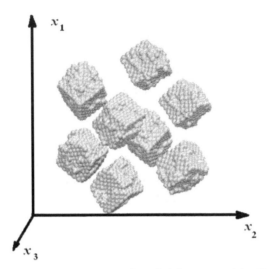

FIGURE 9.12   Nanoparticles system consists of eight nanoparticles NaCl.

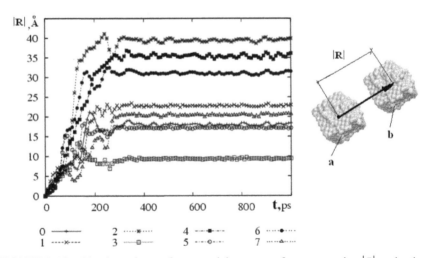

FIGURE 9.13   The dependence of nanoparticle centers of masses motion $|R|$ on the time t; a, b --the nanoparticle positions at time 0 and t, accordingly; 1–8–are the numbers of the nanoparticles.

However, one can observe a number of other situations. Let us consider, for example, the self-organization calculation for the system consisting of 125 cubic nanoparticles, the atomic interaction of which is determined by Morse potential (Figure 9.14).

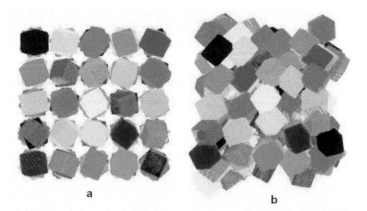

FIGURE 9.14 The positions of the 125 cubic nanoparticles: (a)–initial configuration; (b)—final configuration of nanoparticles.

As you see, the nanoparticles are moving and rotating in the self-organization process forming the structure with minimal potential energy. Let us consider, for example, the calculation of the self-organization of the system consisting of two cubic nanoparticles, the atomic interaction of which is determined by Morse potential [12]. Figure 9.15 displays possible mutual positions of these nanoparticles. The positions, where the principal moment of forces is zero, corresponds to pairs of the nanoparticles 2–3; 3–4; 2–5 (Figure 9.15) and defines the possible positions of their equilibrium.

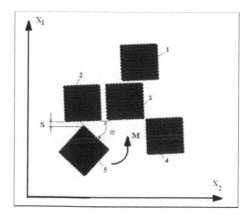

FIGURE 9.15 Characteristic positions of the cubic nanoparticles.

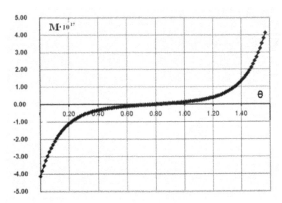

FIGURE 9.16   The dependence of the moment M [Nm] of the interaction force between cubic nanoparticles 1–3 (see in Figure 9.9) on the angle of their relative rotation θ [rad].

Figure 9.16 presents the dependence of the moment of the interaction force between the cubic nanoparticles 1–3 (Figure 9.15) on the angle of their relative rotation. From the plot follows that when the rotation angle of particle 1 relative to particle 3 is $\pi/4$, the force moment of their interaction is zero. At an increase or a decrease in the angle the force moment appears. In the range of $\pi/8 < \theta < 3\pi/4$ the moment is small. The force moment rapidly grows outside of this range. The distance S between the nanoparticles plays a significant role in establishing their equilibrium. If $S > S_0$ (where S_0 is the distance, where the interaction forces of the nanoparticles are zero), then the particles are attracted to one another. In this case, the sign of the moment corresponds to the sign of the angle θ deviation from $\pi/4$. At $S < S_0$ (the repulsion of the nanoparticles), the sign of the moment is opposite to the sign of the angle deviation. In other words, in the first case, the increase of the angle deviation causes the increase of the moment promoting the movement of the nanoelement in the given direction, and in the second case, the angle deviation causes the increase of the moment hindering the movement of the nanoelement in the given direction. Thus, the first case corresponds to the unstable equilibrium of nanoparticles, and the second case–to their stable equilibrium. The potential energy change plots for the system of the interaction of two cubic nanoparticles (Figure 9.17) illustrate the influence of the parameter S. Here, curve 1 corresponds to the condition $S < S_0$ and it has a well-expressed minimum in the $0.3 < \theta < 1.3$ region. At $\theta < 0.3$ and $\theta > 1.3$, the interaction potential energy sharply increases, which leads to the return of the system into the initial equilibrium position. At $S > S_0$ (curves 2–5), the potential energy plot has a maximum at the $\theta = 0$ point, which corresponds to the unstable position.

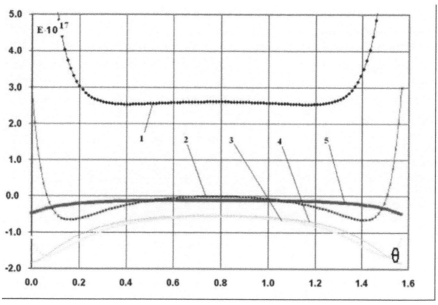

FIGURE 9.17 The plots of the change of the potential energy E [Nm] for the Interaction of two cubic nanoparticles depending on the angle of their relative rotation θ [rad] and the distance between them (positions of the nanoparticles 1–3, Figure 9.9).

The carried-out theoretical analysis is confirmed by the works of the scientists from New Jersey University and California University in Berkeley who experimentally found the self-organization of the cubic micro-particles of plumbum zirconate–titanate (PZT) [28]: the ordered groups of cubic micro-crystals from PZT obtained by hydrothermal synthesis formed a flat layer of particles on the air–water interface, where the particle occupied the more stable position corresponding to position 2–3 in Figure 9.15.

Thus, the analysis of the interaction of two cubic nanoparticles has shown that different variants of their final stationary state of equilibrium are possible, in which the principal vectors of forces and moments are zero. However, there are both stable and unstable stationary states of this system: nanoparticle positions 2–3 are stable, and positions 3–4 and 2–5 have limited stability or they are unstable depending on the distance between the nanoparticles.

Note that for the structures consisting of a large number of nanoparticles, there can be a quantity of stable stationary and unstable forms of equilibrium. Accordingly, the stable and unstable nanostructures of composite materials can appear. The search and analysis of the parameters determining the formation of stable nanosystems is an urgent task.

It is necessary to note that the method offered has restrictions. This is explained by change of the nanoparticles form and accordingly variation of interaction pair potential during nanoparticles coming together at certain conditions.

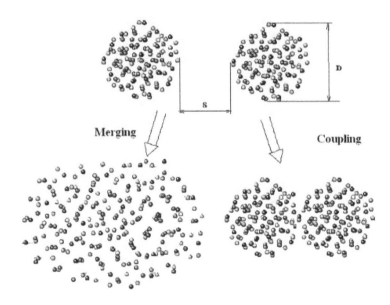

FIGURE 9.18   Different type of nanoparticles connection (merging and coupling).

The merge (accretion [4]) of two or several nanoparticles into a single whole is possible (Figure 9.18). Change of a kind of connection cooperating nanoparticles (merging or coupling in larger particles) depending on its sizes, it is possible to explain on the basis of the analysis of the energy change graph of connection nanoparticles (Figure 9.19). From Figure 9.19 it follows that although with the size increasing of a particle, energy of nanoparticles connection E_{np} grows, its size in comparison with superficial energy E_s of a particle sharply increases at reduction of the sizes of nanoparticles. Hence, for finer particles, energy of connection can appear sufficient for destruction of their configuration under action of a mutual attraction and merging in larger particle.

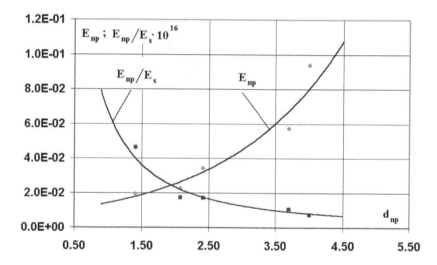

FIGURE 9.19   Change of energy of nanoparticles connection E_{np} [Nm] and E_{np} ratio to superficial energy E_s depending on nanoparticles diameter d [nm]. Points designate the calculated values. ontinuous lines are approximations.

Spatial distribution of particles influences on rate of the forces holding nanostructures formed from several nanoparticles. On Figure 9.20 the chain nanoparticles, formed by coupling of three nanoparticles (from 512 atoms everyone), located in the initial moment on one line. Calculations have shown that, in this case, nanoparticles form a stable chain. Thus, particles practically do not change the form and cooperate on small platforms.

In the same figure (Figure 9.20) the result of connection of three nanoparticles located in the initial moment on a circle and consisting of 256 atoms everyone is submitted. In this case particles incorporate among themselves "densely," contacting on a significant part of the external surface.

Distance between particles at which they are in balance is much less for the particles collected in group ($L^0_{3np} < L^0_{2np}$). It also confirms the graph of forces from which it is visible that the maximal force of an attraction between particles in this case (is designated by a continuous line) is some times more than at an arrangement of particles in a chain (dashed line) $F_{3np} > F_{2np}$.

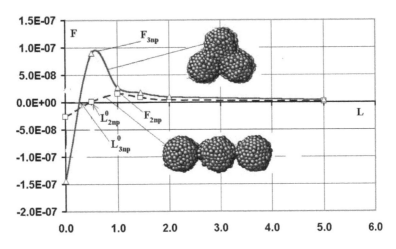

FIGURE 9.20 Change of force F [N] of three nanoparticles interaction, consisting of 512 atoms everyone, and connected among themselves on a line and on the beams missing under a corner of 120 degrees, accordingly, depending on distance between them L [nm].

Experimental investigation of the spatial structures formed by nanoparticles [4] confirms that nanoparticles gather to compact objects. Thus, the internal nuclear structure of the connections area of nanoparticles considerably differs from structure of a free nanoparticle.

Nanoelements' kind of interaction depends strongly on the temperature. Figure 9.21 shows the picture of the interaction of nanoparticles at different temperatures It is seen that with increasing temperature the interaction of changes in sequence occurs (Figure 9.22): coupling (1,2) and merging (3,4). With further increase in temperature the nanoparticles dispersed.

FIGURE 9.21   Change of nanoparticles connection at increase in temperature.

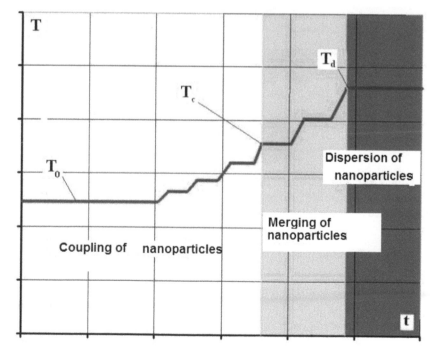

FIGURE 9.22   Curve of temperature change.

Finally, we will consider problems of dynamics of nanoparticles. The analysis of interaction of nanoparticles among themselves also allows to draw a conclusion on an essential role in this process of energy of initial movement of particles. Various processes at interaction of the nanoparticles, moving with different speed, are observed: the processes of agglomerate formation, formation of larger particles at merge of the smaller size particles, absorption by large particles of the smaller ones, and dispersion of particles on separate smaller ones or atoms.

For example, in Figure 9.23, the interactions of two particles are moving towards each other with different speed are shown. At small speed of moving is formed steady agglomerate (Figure 9.23).

In Figure 9.23 (left), interaction of two particles moving towards each other with the large speed is shown. It is visible that steady formation in this case is not appearing and the particles collapse.

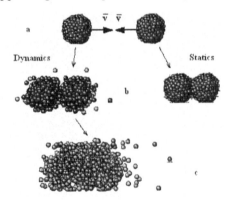

FIGURE 9.23 Pictures of dynamic interaction of two nanoparticles: (a) an initial configuration nanoparticles; (b) nanoparticles at dynamic interaction, (c) the "cloud" of atoms formed because of dynamic destruction two nanoparticles.

Nature of interaction of nanoparticles, along with speed of their movement, essentially depends on a ratio of their sizes. In Figure 9.24, pictures of interaction of two nanoparticles of zinc of the different size are presented, as well as in the earlier case of a nanoparticle move from initial situation (1) towards each other. At small initial speed, nanoparticles incorporate at contact and form a steady conglomerate (2).

FIGURE 9.24   Pictures of interaction of two nanoparticles of zinc: 1 – initial configuration of nanoparticles; 2 – connection of nanoparticles, 3,4 – absorption by a large nanoparticle of a particle of the smaller size, 5 –destruction of nanoparticles at blow.

At increase in speed of movement larger nanoparticle absorbs smaller, and the uniform nanoparticle (3, 4) is formed. At further increase in speed of movement of nanoparticles, owing to blow, the smaller particle intensively takes root in big and destroys it.

The given examples show that use of dynamic processes of pressing for formation of nanocomposites demands a right choice of a mode of the loading providing integrity of nanoparticles. At big energy of dynamic loading, instead of a nanocomposite with dispersion corresponding to the initial size of nanoparticles, the nanocomposite with much larger grain that will essentially change properties of a composite can be obtained.

9.8   CONCLUSIONS

In conclusion, the following basic regularities of the nanoparticle formation and self-organization should be noted.

1. The existence of several types of the forms and structures of nanoparticles is possible depending on the thermodynamic conditions.
2. The absence of the crystal nucleus in small nanoparticles.
3. The formation of a single (ideal) crystal nucleus defects on the nucleus surface connected to the amorphous shell.
4. The formation of the polycrystal nucleus with defects distributed among the crystal grains with low atomic density and change interatomic distances. In addition, the grain boundaries are non-equilibrium and contain a great number of grain-boundary defects.
5. When there is an increase in sizes, the structure of nanoparticles is changing from amorphous to roentgen-amorphous and then into the crystalline structure.
6. The formation of the defect structures of different types on the boundaries and the surface of a nanoparticle.
7. The nanoparticle transition from the globe-shaped to the crystal-like shape.
8. The formation of the "regular" and "irregular" shapes of nanoparticles depending on the number of atoms forming the nanoparticle and the relaxation conditions (the rate of cooling).
9. The structure of a nanoparticle is strained because of the different distances between the atoms inside the nanoparticle and in its surface layers.
10. The systems of nanoparticles can form stable and unstable nanostructures.

ACKNOWLEDGMENTS

This work was carried out with financial support from the Research Program of the Ural Branch of the Russian Academy of Sciences: the projects 12-P-12-2010, 12-C-1-1004, 12-T-1-1009, and was supported by the grants of Russian Foundation for Basic Research (RFFI) 04-01-96017-r2004ural_a, 05-08-50090-a, 07-01-96015-r_ural_a, 08-08-12082-ofi, and 11-03-00571-a. The author is grateful to his young colleagues Dr. A.A. Vakhrushev and Dr. A.Yu. Fedotov for active participation in the development of the software complex, calculations, and analyzing of numerous calculation results. The calculations were performed at the Joint Supercomputer Center of the Russian Academy of Sciences.

KEYWORDS

- **Hypocrystalline phases**
- **Nanocomposites**
- **Nanostructural elements**

REFERENCES

1. Qing-Qing, Ni.; Yaqin, Fu.; Masaharu Iwamoto, *"Evaluation of Elastic Modulus of Nano Particles in PMMA/Silica Nanocomposites"*. *J. Soc. Mater. Sci. Japan.* **2004,** *53(9),* 956–961.
2. Ruoff, R. S.; Pugno, Nicola M.; *"Strength of nanostructures"*. Mechanics of the 21st Century. Proceeding of the 21-th international congress of theoretical and applied mechanics. Warsaw: Springer; **2004,** 303–311 p.
3. Diao, J.; Gall, K.; Dunn, M. L.; *"Atomistic simulation of the structure and elastic properties of gold nanowires"*. *J. Mech. Phys. Solids.* **2004,** *52(9),* 1935–1962.
4. Dingreville, R.; Qu, J.; and Cherkaoui, M.; *"Surface free energy and its effect on the elastic behavior of nano-sized particles, wires and films"*. *J. Mech. Phys. Solids.* **2004,** *53(8),* 1827–1854.
5. Duan, H. L.; Wang, J.; Huang, Z. P.; Karihaloo, B. L.; *"Size-dependent effective elastic constants of solids containing nano-inhomogeneities with interface stress"*. *J. Mech. Phys. Solids.* **2005,** *53(7),* 1574–1596.
6. Gusev, A. I.; and Rempel, A. A.; *" Nanocrystalline materials"* Moscow : Physical Mathematical Literature; **2001,** (in Russian).
7. Hoare, M. R.; *"* Structure and dynamics of simple microclusters". *Ach. Chem. Phys.* **1987,** *40,* 49–135.
8. *Brooks, B. R.; Bruccoleri, R. E.; Olafson, B. D.; States, D. J.; Swaminathan, S.; and Karplus, M.; "CHARMM: A program for macromolecular energy minimization, and dynamics calculations". J. Comput. Chem.* **1983,** *4(2), 187–217.*
9. Friedlander, S. K.; *"Polymer-like behavior of inorganic nanoparticle chain aggregates ."* *J. Nanoparticle Res.* **1999,** *1,* 9–15.
10. Grzegorczyk, M.; Rybaczuk, M.; and Maruszewski, K.; *" Ballistic aggregation: an alternative approach to modeling of silica sol–gel structures"* Chaos, Solitons and Fractals, 19, 1003–11, **2004.**
11. Shevchenko, E. V.; Talapin, D. V.; Kotov, N. A.; O'Brien, S.; and Murray, C. B.; *"Structural diversity in binary nanoparticle superlattices" Nature Lett.* **2006,** *439,* 55–59.

12. Kang, Z. C.; and Wang, Z. L.; *"On Accretion of nanosize carbon spheres"*. *J. Phys. Chem.* **1996,** *100,* 5163–65.
13. Melikhov, I. V.; and Bozhevol'nov, V. E.; *"Variability and self-organization in nano-systems"*. *J. Nanoparticle Res.* **2003,** *5,* 465–72.
14. Kim, D.; and Lu, W.; *"Self-organized nanostructures in multi-phase epilayers"* *Nano-technol.* **2004,** *15,* 667–74.
15. Kurt, E.; Geckeler, *"Novel supermolecular nanomaterials: from design to reality"*, Proceeding of the 12 Annual International Conference on Composites/Nano Engineering, Tenerife, Spain, August 1–6, CD Rom Edition, **2005**.
16. Vakhrouchev, A. V.; and Lipanov, A. M.; *"A numerical analysis of the rupture of powder materials under the power impact influence"*. *Computer & Structures.* **1992,** *1/2(44),* 481–86.
17. A.V. Vakhrouchev, *"Modeling of static and dynamic processes of nanoparticles inter-action"*, CD-ROM Proceeding of the 21-th international congress of theoretical and applied mechanics, ID12054, Warsaw, Poland, **2004**.
18. Vakhrouchev, A. V.; *"Simulation of nanoparticles interaction"*, Proceeding of the 12 Annual International Conference on Composites/Nano Engineering, Tenerife, Spain, August 1–6, CD Rom Edition, **2005**.
19. Vakhrouchev, A. V.; *"Simulation of nano-elements interactions and self-assembling"*. *Modeling and Simulation in Materials Science and Engineering.* **2006,** *14,* 975–991
20. Vakhrouchev, A. V.; and Lipanov, A. M.; *Numerical Analysis of the Atomic structure and Shape of Metal Nanoparticles Computational Mathematics and Mathematical Physics.* **2007,** *47(10),* 1702–1711.
21. Vakhrouchev, A. V.; Modelling of the Process of Formation and Use of Powder Nano-composites Composites with Micro and Nano-Structures. Computational Modeling and Experiments. Computational Methods in Applied Sciences Series. Barcelona, Spain: Springer Science. **2008,** *9,* 107–136.
22. Vakhrouchev, A. V.; Modeling of the nanosystems formation by the molecular dynam-ics, mesodynamics and continuum mechanics methods. *Multidiscipline Modeling in Material and Structures.* **2009,** *5(2),* 99–118.
23. Vakhrouchev, A. V.; Theoretical Bases of Nanotechnology Application to Thermal Engines and Equipment. Izhevsk: Institute of Applied Mechanics, Ural Branch of the Russian Academy of Sciences; **2008**, 212 p. (in Russian).
24. Alikin, V. N.; Vakhrouchev, A. V.; Golubchikov, V. B.; Lipanov, A. M.; Serebrennikov, S. Y.; Development and Investigation of the Aerosol Nanotechnology. Moscow: Mash-inostroenie, **2010**. 196 p. (in Russian)
25. Heerman, W. D.; Computer Simulation Methods in Theoretical Physics. Berlin: Springer-Verlag; **1986.**
Verlet, L.; *"Computer "experiments" on classical fluids I. Thermo dynamical properties of Lennard-Jones molecules"*, Phys. Rev., 159, 98–103, **1967**.
26. Korn, G. A.; and Korn, M. T.; Mathematical Handbook. New York: McGraw-Hill Book Company; **1968**.
27. Self-organizing of microparticles piezoelectric materials, News of chemistry, date-news.php.htm.

CHAPTER 10

TRENDS IN OZONATION OF POLYMER COMPOUNDS

M. P. ANACHKOV, S. K. RAKOVSKY, and G. E. ZAIKOV

CONTENTS

10.1   INTRODUCTION

The interest in the reaction of ozone with polydienes is mainly due to the problems of ozone degradation of rubber materials [1–4] and the application of this reaction to the elucidation of the structures of elastomers [5–8]. It is also associated with the possibilities of preparing bifunctional oligomers by partial ozonolysis of some unsaturated polymers [9–12]. Usually, the interpretation of experimental results are based on a simplified scheme of Criegee's mechanism of C=C-double bond ozonolysis, explaining only the formation of the basic product, ozonides [13, 14].

The reactions of ozone with 1,4-*cis*-polybutadiene Diene 35 NFA having the linking of the butadiene units in the rubber macromolecules as in what follows: 1,4-*cis* (47%), 1,4-*trans* (42%), 1,2-*cis* (11%), 1,4-*cis*-polyisoprene (Carom IR 2200), 1,4-*trans*-polychloroprene (Denka M 40), and 1,4-*trans*-polyisoprene have been investigated in CCl_4 solutions. The changes of the viscosity of the polymer solutions during the ozonolysis have been characterized by the number of chain scissions per molecule of reacted ozone (φ). The influence of the conditions of mass-transfer of the reagents in a bubble reactor on the respective φ values has been discussed. The basic functional groups—products from the rubbers ozonolysis have been identified and quantitatively characterized by means of IR-spectroscopy and ^{1}H-NMR spectroscopy. A reaction mechanism, which explains the formation of all identified functional groups, has been proposed. It has been shown that the basic route of the reaction of ozone with elastomer double bonds, the formation of normal ozonides, does not lead directly to a decrease in the molecular mass of the elastomer macromolecules because the respective 1,2,4-trioxolanes are relatively stable at ambient temperature. The most favorable conditions for ozone degradation emerge when the cage interaction between Criegee intermediates and respective carbonyl groups does not proceed. The amounts of measured different carbonyl groups have been used as an alternative way for evaluation of the intensity and efficiency of the ozone degradation. The thermal decomposition of partially ozonized diene rubbers has been investigated by Differential Scanning Calorimetry DSC. The respective values of the enthalpy, the activation energy, and the reaction order of the 1,2,4-trioxolanes have been determined.

In most cases, quantitative data on the functional groups formed during the reaction are missing [15–18]. At the same time alternative conversion routes of Criegee's intermediates, which lead to the formation of carbonyl compounds and some other so-called "anomalous products" of the ozonolysis, are of great importance for clarifying the overall reaction mechanism [19–21]. The mechanism of ozone degradation of rubbers is also connected with the non-ozonide routes of the reaction, because the formation of the basic product of ozonolysis, normal ozonide, does not cause any chain scission and/or macromolecule cross-linking [22].

In this article, the changes in the molecular mass of different types of diene rubbers during their partial ozonolysis in solution have been investigated. By means of IR and ^{1}H-NMR spectroscopy ozonolysis products of the elastomers have been studied. The effects of the nature of the double bond substituents and its configuration on the degradation mechanism have been considered. By using differential scanning calorimetry the thermal decomposition of the functional groups of peroxide type has also been investigated.

10.2   EXPERIMENTAL

10.2.1   MATERIALS

Commercial samples of 1,4-*cis*-polybutadiene (SKD, E-BR); polybutadiene (Diene 35 NFA, BR); 1,4-*cis*-polyisoprene (Carom IR 2200, E-IR), and polychloroprene (Denka M 40, PCh) were used in the experiments (Table 10.1). The 1,4-*trans*-polyisoprene samples were supplied by Prof. A. A. Popov, Institute of Chemical Physics, Russian Academy of Sciences. All rubbers were purified by three-fold precipitation from CCl_4 solutions in excess of methanol. The aforementioned elastomer structures were confirmed by means of ^{1}H-NMR spectroscopy. *Ozone* was prepared by passing oxygen flow through a 4–9 kV electric discharge.

10.2.2   OZONATION OF THE ELASTOMER SOLUTIONS

The ozonolysis of elastomers was performed by passing an ozone—oxygen gaseous mixture at a flow rate of $v = 1.6 \quad 10^{-3} \pm 0.1$ l/s through a bubbling reactor, containing 10–15 ml of polymer solution (0.5–1 g in CCl_4) at 293 K. Ozone concentrations in the gas phase at the reactor inlet ($[O_3]_i$) and outlet ($[O_3]_u$) were measured spectrophotometrically at 254 nm

[23]. The amount of consumed ozone (G, mole) was calculated by the Eq. (10.1):

$$G = v([O_3]_I - [O_3]_u)t \qquad (10.1),$$

where t is the ozonation time (s). The degree of conversion of the double C=C bonds was determined on the basis of the amount of reacted ozone and the reaction stoichiometry [23].

TABLE 10.1   Some characteristics of polydiene samples

Elastomer	Monomeric unit	Unsaturation degree (%)	1,4-*cis* (%)	1,4-*trans* (%)	1,2- (%)	3,4- (%)	$M_v \times 10^{-3}$	n
SKD	–CH=CH-	95–98	87–93	3–8	3–5	–	454	2.1
Diene 35 NFA	–CH=CH-	97	47	42	11		298	2.63
Carom IR 2200	–C(CH$_3$)=CH-	94–98	94–97	2–4	–	1–2	380	2.0
1,4-*Trans* PI	–C(CH$_3$)=CH-	95–97		95–97			310	2.3
Denka M40	–C(Cl)=C-	94–98	5	94	–	–	180	1.8

Note: M_v is the average molecular weight, determined viscosimetrically from equation $[\eta] = k \times M_v^{a'}$, where $[\eta] = (\eta_1/C)(1+0.333\eta_1)$, $\eta_1 = \eta_{rel} - 1$, η_{rel} is the intrinsic viscosity; C is the solution concentration; $k = 1.4 \times 10^{-4}$ —Staudinger's constant, and $\alpha' = 0.5 - 1.5$—constant depending on the rubber type, being one for natural rubber; M_v M_w; $n = M_w/M_n$, where M_w and M_n are the average weight and number average molecular mass, respectively [22].

10.3   RESULTS AND DISCUSSION

Florry [24] has shown that the reactivity of the functional groups in the polymer molecule does not depend on its length. It is also known that some reactions of the polymers proceed more slowly when compared with their low molecular analogs (catalytic hydrogenation). The folded or unfolded forms of the macromolecules provide various conditions for contact of the reagents with the reacting parts [4, 25]. By using the modified version of this principle [26], it is possible to explain the proceeding of reactions without any specific interactions between the adjacent C=C bonds and the absence of diffusion limitations. The study of the mass-molecular distribution (MMD) is a very sensitive method for establishing the correlation between molecular weight (M_w) and the reactivity.

The theory predicts that the properties of the system, polymer—solvent can be described by the parameter of so-called globe swelling (γ), which defines the free energy (F) of the system and thus the rate constant of the reaction. For a reversible reaction, i.e., polymerization—depolymerization, the dependence of the rate constant of the chain length growing on the molecular weight is expressed by the following equation:

$$\ln k_{pj}/k_{p\infty} = -\text{const. } (5\gamma - 3/\gamma).(d\gamma/dM).M_0 \qquad (10.2)$$

where M_0 is the molecular weight of the studied sample and k_p is the rate constant for infinitely long macromolecules. A good correlation between the theoretical and experimental data for polystyrene solutions in benzene has been found in Ref. [27].

The study of the polymer degradation is complicated by their structural peculiarities on molecular and supramolecular level and diffusion effects. It is difficult to find simple model reactions for clarification of particular properties and for the express examination of the proposed assumptions. An exception, in this respect, is the ozone reaction with C=C bonds, whose mechanism has been intensively studied and could be successfully applied upon ozonolysis of polymeric materials [28].

Table 10.2 summarizes the rate constants of the ozone reactions with some conventional elastomers and polymers and their low molecular analogs, synthesized by us. It is seen that the reactivities of elastomers and polymers and their corresponding low molecular analogs, as it is demonstrated by their rate constants, are quite similar, thus suggesting similar mechanisms of their reaction with ozone. This statement is also confirmed by: (1) the dependence of k on the inductive properties of substituents: for example k of polychloroprene is higher than that of vinylchloride due to the presence of two donor substituents and (2) the dependence of k on the configuration of the C=C bond in *trans*-isomer (gutta-percha) and *cis*-isomer (natural rubber).

TABLE 10.2 Rate constants of ozone reactions with polymers and low molecular analogs in CCl_4, 20°C

Compound	M.W.	$k \times 10^{-4}$, M^{-1} s^{-1}
Polychloroprene	8×10^5	0.42 ± 0.1
Vinylchloride	62.45	0.18
2-Bromopropene	121	0.28 ± 0.05
Polybutadiene	3.3×10^5	6.0 ± 1
Cyclododecatriene-1,5,9	162	35 ± 10
Poly(Butadiene-Co-Styrene)	8×10^4	6 ± 1
Gutta-percha	3×10^4	27 ± 5
Natural Rubber	1×10^6	44 ± 10
2-Me-Pentene-2	85	35 ± 10
Squalene	410	74 ± 15
Polystyrene	5×10^5	0.3×10^{-4}
Cumene	120	0.6×10^{-4}
Polyisobutylene	1.7×10^5	0.02×10^{-4}
Cyclohexane	84	0.01×10^{-4}

It has been found out that the effects, related either to the change in the macromolecule length or to the folding degree, do not affect the ozonolysis in solution. Probably this is due to the fact that the reaction is carried out in elastomeric solutions, in which the macromolecules are able to do free intramolecular movements and they do not react with adjacent macromolecules. Moreover, the rate of macromolecules reorganization is probably higher than the rate of their reaction with ozone as the experiment does not provide any evidence for the effects of the change in the parameters mentioned above [29].

However, it should be noted that k values of the elastomers are about 2–6 times lower than those of the low molecular analogs.

The accuracy of activation energy (E_a) determination does not allow to estimate the contribution of the two parameters: pre-exponential factor (A) or E_a for the decrease in k. If we assume that the mechanism of ozone reaction with monomers and elasomers is similar, i.e., the reactions are isokinetic, then $A_{mon} = A_{pol}$. At $k_{mon}/k_{pol} = 2 \div 6$ the difference in E_a at 20°C will be 0.5–1.0 kcal/mol. At the low experimental values of E_a, these differences will become commensurable and thus the determination of E_a is not sufficiently accurate. In this case two assumptions could be made which can give a reasonable explanation for the lower values of k_{pol}: (1) the reorientation of the macromolecules is a slower process than that of olefins, which would results in $A_{pol} < A_{mon}$ and (2) the addition of ozone to C=C bonds is accompanied by the rehybridization of the C-atoms from sp^2 to sp^3 and the movements of the polymer susbstituents during the formation of activated complex will be more restricted than those in olefins, mainly because of their greater molecular mass and sizes.

This will ultimately result in decrease of the rate constant. Table 10.2 shows some examples of ozonolysis of saturated polymers—polystyrene and polyisobutylene. These reactions take place not through the mechanism of ozone reaction with the double bonds but through a hidden radical mechanism with rate constants of four to five orders of magnitude lower.

10.3.1 POLYBUTADIENES

Because of the high viscosity and high value of rate constants, the reaction takes place either in the diffusion or in the mixed region. To obtain correct kinetic data we have used the theory of boundary surface [30]:

$$[O_3] = \alpha[O_3]_0 \times \exp[-\delta(kcD)^{1/2}], \qquad (10.3)$$

where $[O_3]$ is the ozone concentration at a distance δ, α the Henry's coefficient, $[O_3]_0$ the equilibrium ozone concentration in the gas phase at the reactor inlet, δ the penetration depth of ozone from the interphase surface [22], k the rate constant of the ozone reaction with double bonds, c the concentration of the monomeric units, and D is the diffusion coefficient of ozone in the liquid phase.

It was found out that the relative viscosity decreases exponentially upon ozonation of SKD solutions (Figure 10.1). As the viscosity is proportional to the molecular weight it follows that the polydiene consumption should be described by first or pseudo-first order kinetics.

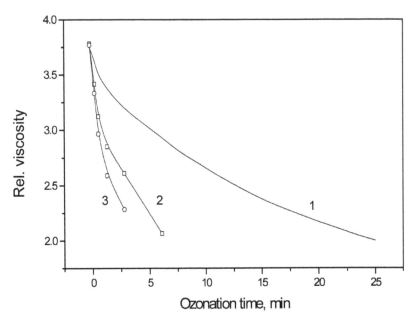

FIGURE 10.1   Dependence of the relative viscosity (η_{rel}) of SKD solutions (0.6 g in 100 ml CCl$_4$) on reaction time at ozone concentrations of (1)—1 × 10^{-5} M; (2)—4.5 × 10^{-5} M, and (3)—8.25 × 10^{-5} M.

The value of ϕ, corresponding to the number of degraded polymeric molecules per one absorbed ozone molecule can be used to calculate the degradation efficiency. The value of this parameter (ϕ) may be estimated using the following equation:

$$\phi = 0.5 \left[(M_{vt})^{-1} - (M_{v0})^{-1} \right] P/G, \tag{10.4}$$

where M_{vt} is the molecular weight at time moment T, M_{v0} the initial molecular weight, P the polymer amount, and G is the -amount of consumed ozone.

The dependence of ϕ on G is a straight line for a given reactor and it depends on the hydrodynamic conditions in the reactor. It is seen from Figure 10.2 that the ϕ values are increasing linearly with the reaction time and decreasing with increase in ozone concentration. The corresponding dependences for Carom IR 2200 and Denka M40 ozonolysis are similar. The ϕ values for G ® 0 were used to avoid the effect of hydrodynamic factors on them.

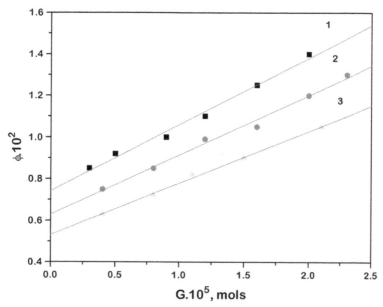

FIGURE 10.2   Dependence of ϕ on G for SKD (0.6/100) at various ozone concentrations: (1)—1×10^{-5} M, (2)—4.5×10^{-5} M, and (3)—8.25×10^{-5} M.

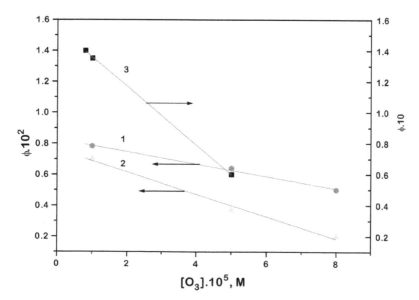

FIGURE 10.3   Dependence of ϕ on ozone concentration for elastomer solutions: —(1) SKD (0.6/100), (2) Carom IR 2200 (0.6/100), and (3) Denka M40 (1/100).

The values of ϕ found for SKD, Carom IR 2200, and Denka M40 at $[O_3]=1 \times 10^{-5}$ M amount to 0.7×10^{-2}, 0.78×10^{-2}, and 0.14, respectively, and the slopes are -40, -70, and 200 M^{-1}, respectively. Substituting with the known values for the parameters in Equation (3) we have obtained δ within the range of $1 \times 10^{-3} - 2 \times 10^{-4}$ cm, which indicates that the reaction is taking place in the volume around the bubbles, and hence in the diffusion region.

The ozonolysis of polydienes in solutions is described by the Criegee's mechanism. The C=C bonds in the macromolecules are isolated as they are separated by three simple C–C bonds. According to the classical concepts, the C=C bonds configuration and the electronic properties of the groups bound to them, also affect the polymer reactivity; similarly they do this in case of the low molecular olefins. The only difference is that the polymer substituents at the C=C bonds are less mobile, which influences the sp^2–sp^3 transition and the ozonides formation. In the first stage, when primary ozonides (POs) (Scheme 1, reaction 1) are formed, the lower mobility of the polymer substituents requires higher transition energy, the rate being respectively lower, compared to that with low molecular olefins and the existing strain accelerates the PO decomposition to zwitterion and carbonyl compound.

The lower mobility of the polymer parts impedes the further ozonide formation and causes the zwitterion to leave the cage and pass into the volume, which in turn accelerates the degradation process. The latter is associated with either its monomolecular decomposition or its interaction with low molecular components in the reaction mixture. The efficiency of degradation is determined by the C=C bonds location in the macromolecule, for example, at C=C bond location from the macromolecule center to its end, is in the range from 2 to 1.

$$M_1 = (1/\gamma)M_0, \tag{10.5}$$

where $M_2 = M_0 - M_1$; $1 \leq \gamma \leq 2$ is the coefficient pointing the C=C-bond location; M_0, M_1, and M_2 are the molecular weights of the initial macromolecule and of the two degraded polymer parts, respectively. At $\gamma = 2$, i.e., when the broken C=C bond is located in the macromolecule center, the values of M_1 and M_2 will be exactly equal to $M_0/2$, at $\gamma \circledR 1$, i.e., at terminal C=C bond in the polymer chain, the value of M_1 will be approximated to M_0 and thus the value of M_2 will be practically insignificant. For example,

M_2 may be 50–1,000, which is three to four orders of magnitude less than that of the macromolecule and in fact degradation process will not occur.

The viscosimetric determination of the molecular weight, which we have applied in our experiments, has accuracy of ±5 per cent and does not allow the differentiation of molecular weights of 22,700, 19,000, and 9,000 for the corresponding types of rubbers. This suggests that the cleavage of C=C bonds, located at distances of 420, 280, and 100 units from the macromolecule end, would not affect the measured molecular weight.

Since the reaction of elastomers ozonolysis proceeds either in the diffusion or in diffusion-kinetic region, at low conversions each new gas bubble in the reactor would react with a new volume of the solution. On the other hand, the reaction volume is a sum of the liquid layers surrounding each bubble. It is known that the depth of the penetration from the gas phase into the liquid phase is not proportional to the gas concentration and thus the rise of ozone concentration would increase the reaction volume to a considerably smaller extent than the ozone concentration.

This leads to the occurrence of the following process: intensive degradation processes take place in the micro-volume around the bubble and one macromolecule can be degraded to many fragments, whereas the macromolecules out of this volume, which is much greater, may not be changed at all. Consequently with increase in ozone concentration, one may expect a reduction of coefficient MMD and increase in the oligomeric phase content. This will result in apparent decrease of ϕ in case of the viscosimetric measurements. The discussion above enables the correct interpretation of the data in Figure 10.3.

In the spectra of the ozonized polybutadienes the appearance of bands at 1,111 and 1,735 cm^{-1}, which are characteristic for ozonide and aldehyde groups, respectively, is observed [22, 31]. It was found that the integral intensity of ozonide peak in the 1,4-*cis*-polybutadiene Emulsion Butadiene Rubber (E-BR) spectrum is greater and that of the aldehyde is considerably smaller in comparison with the respective peaks in the Diene 35 NFA (BR) spectrum at one and the same ozone conversion degree of the double bonds. The differences in the aldehyde yields indicate that, according to IR-analysis, the degradation efficiency of the BR solutions is greater.

The ^{1}H-NMR spectroscopy provides much more opportunities for identification and quantitative determination of functional groups, formed during ozonolysis of polybutadienes [32]. Figure 10.4 shows spectra of ozonized E-BR. The signals of the ozonolysis products are decoded in Table 10.3 on the basis of Figure 10.5. The ozonide—aldehyde ratio, determined from NMR spectra, was 89:11 and 73:27 for E-BR and BR, respectively. The peak at 2.81 ppm is present only in the spectra of ozonized Diene 35 NFA. It is usually associated with the occurrence of epoxide groups [33]. The integrated intensity of that signal as compared to the signal of aldehyde protons at 9.70–9.79 ppm was about 10 per cent. Similar signal at 2.75 ppm has been registered in the spectra of ozonized butadiene-nitrile rubbers, where the 1,4-*trans* double bonds are dominant [31].

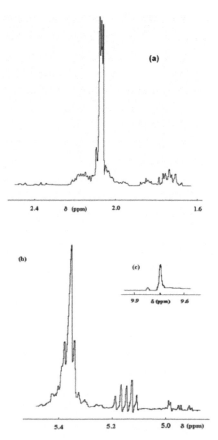

FIGURE 10.4   (a,b,c) 1H-250 MHz NMR spectra of E-BR solutions (0.89 g/100 ml CCl$_4$) ozonized to 18 per cent conversion of the double bonds (external standard TMS; digital resolution 0.4 Hz, 20°C).

According to Refs. [2, 10] two isomeric forms of 1,2,4-trioxolanes exist. The ratio between them is a function of the double bond stereochemistry, steric effect of the substituents, and the conditions of ozonolysis. It was found out only on the low molecular weight alkenes [19, 21]. The ¹H-NMR spectroscopy is the most powerful method for determination of the *cis/trans* ratio of ozonides (in the case of polymers it is practically the only one method that can be applied). The measuring is based on the differences in the chemical shifts of the methine protons of the two isomers: the respective signal of the *cis* form appears in lower field as compared to the *trans* one [19, 21].

The multiplet on Figure 10.3 in the area of 5.1–5.18 ppm could be interpreted as a result of partial overlapping of triplets of *trans-* and *cis-*ozonides: 5.12 ppm (t, J 5 Hz, 2H) and 5.16 ppm (t, J 5 Hz, 2H), respectively. It is interesting to note that the *cis/trans* ratio of the E-BR 1,2,4-trioxolanes is practically equal to that obtained from *cis-*3-hexene [19, 34]. The resolution of the respective BR spectrum does not allow consideration in detail of the multiplicity of the signals at 5.10 and 5.15 ppm. In this case, the area of the signals is widened, most probably due to the presence of ozonide signals of the 1,2-monomer units [20].

The basic route of the reaction, the formation of normal ozonides, does not lead directly to a decrease in the molecular mass of the elastomer macromolecules, because the respective 1,2,4-trioxolanes are relatively stable at ambient temperature (by analogy with Scheme 1 of polyisoprenes, see as in what follows) [31, 32]. The most favorable conditions for ozone degradation emerge when the cage interaction (Scheme 1, reaction 3) does not proceed. Therefore, the higher the ozonide yield the lower the intensity of ozone degradation of the polybutadienes and vice versa. As it was already determined the ozonide yields for the 1,4-*cis-* and 1,2- monomer units are close to 83–90 per cent, whereas that for the 1,4-*trans* units is about 50 per cent.

The amount of aldehyde groups is usually used for evaluation of the intensity and efficiency (number of chain scissions per molecule of reacted ozone) of ozone degradation of elastomers. In this case, it should be taken into account that the dominant route of degradation leads to the formation of 1 mole of aldehyde from 1 mole of ozone [32].

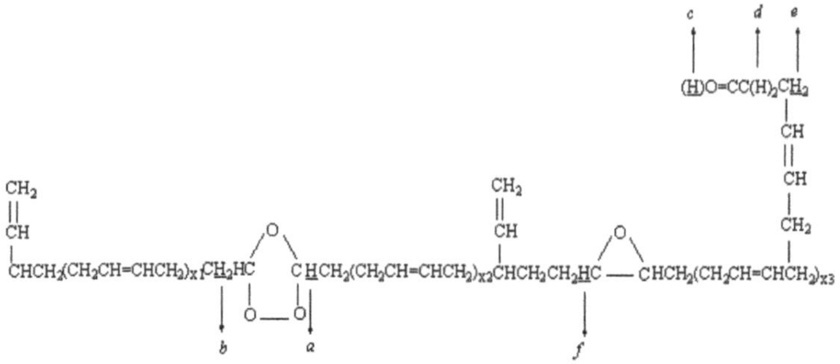

where x_1, x_2, x_3 = 1, 2, 3......n

FIGURE 10.5   Selection of protons with characteristic signals in the ^{1}H-250 MHz NMR spectra of partially ozonized polybutadiene macromolecules.

10.3.2   POLYISOPRENES

The positive inductive effect of the methyl group in polyisoprene enhances the rate of ozone addition to the double bonds from 6 10^4 for SKD to 4 10^5 $M^{-1}s^{-1}$ for Carom IR 2200. The infrared spectra of ozonized 1,4-*trans*-polyisoprene (Z-IR) show two intense bands at 1,100 and 1,725 cm^{-1}, characteristic of ozonide and keto groups, respectively [19, 35]. These spectra are identical with the well-known spectra of 1,4-*cis*-polyisoprenes (E-IR), as far as ozonide and carbonyl bands are concerned [22, 36]. It was found that the integrated intensity of the peak at 1,100 cm^{-1} in the E-IR and Z-IR spectra is equal for one and the same amount of reacted ozone. By analogy with the peak at 1110 cm^{-1}, the integral intensity of the peak at 1,725 cm^{-1} is also one and the same. The latter show that, according to the infrared spectra, the degradation efficiencies of E-IR and Z-IR with respect to the amount of consumed ozone do not practically differ.

TABLE 10.3 Assignment of the signals in the ¹H-NMR spectra of partially ozonized E-BR and BR rubbers

Assignment of the signals	Chemical shifts (ppm)		Literature
(according Figure 10.4)	E-BR	BRA	
a	5.10–5.20	5.05–5.18	[19, 34]
	max 5.12, 5.16	max 5.10, 5.15	
b	1.67–1.79	1.66–1.80	[7, 19]
	max 1.72, 1.76	max 1.73	
c	9.75	9.74	[33]
d	2.42–2.54	2.42–2.54	[33]
	max 2.47	max 2.50	
e	2.27–2.42	2.27–2.42	[33]
	max 2.35	max 2.35	
f		max 2.81	[33]

The ¹H-NMR spectroscopy affords much more opportunities for identification and quantitative determination of functional groups formed on ozonolysis of polyisoprenes. Figure 10.6 shows spectra of non-ozonized and ozonized E-IR. Changes in the spectra of ozonized elastomers are decoded in Table 10.4 on the basis of Figure 10.7. It is seen that besides ketones, aldehydes are also formed as a result of ozonolysis. A comparison between methylene and methine proton signals of the non-ozonized polyisoprenes and the corresponding ozonide signals indicates a considerable overlap in the ranges 1.60–1.90 and 4.5–5.5 ppm. A signal overlap is also registered in the 2.00–2.20 ppm region, characteristic of methyl protons of keto groups. Because of the reasons mentioned above, the ozonides and aldehydes were quantified by using the integrated intensity of the signals at 1.40 (a) and 9.70–9.79 (g) ppm, respectively.

Ketone amounts were determined as a difference between the total intensity of the methylene signals from aldehydes and ketones, 2.40–2.60 and 2.35–2.60 ppm, respectively, and the doubled intensity of the aldehyde signal at 9.70–9.79 ppm. Thus, the obtained ozonide—ketone—aldehyde ratio was 40:37:23 and 42:39:19 for 1,4-*cis*-polyisoprene and 1,4-*trans*-polyisoprene, respectively. The peak at 2.73 ppm, present in the spectra of both ozonized elastomers, is associated with the occurrence of epoxide groups [33, 37]. The integrated intensity of that signal as compared to the signal at 9.70–9.79 ppm was 21 and 15 per cent for E-IR and Z-IR, respectively.

Current ideas about the mechanism of C=C double bond ozonolysis in solution are summarized in Schemes 1 and 2 [19, 21, 34]. As a result of the decomposition of the initial reaction product, PO, zwitterionic species is formed, termed as Criegee's intermediates or carbonyl oxides (hereafter referred to as CI) (Scheme 1, reactions 2 and 2'). Two intermediates are formed from asymmetric olefins: monosubstituted CI (MCI) and disubstituted CI (DCI), if their *syn* and *anti* stereoisomers are not taken into account.

It is known that carbonyl oxides are predominantly formed at carbon atoms with electron-donating substituents [19]. Excellent correlations of the regioselectivities of MO fragmentation with electron donation by substituents (as measured by Hammett and Taft parameters) have been obtained, consistent with the effects expected for stabilization of a zwiterionic carbonyl oxide [20]. According to Ref. [23], for polyisoprenes the ratio between the two intermediates, DCI and MCI, is 64:36.

Ozonides are the basic product of polyisoprene ozonolysis in nonparticipating solvents. It is known that the dominant part of ozonides is formed through the interaction between CI and the corresponding carbonyl group, which originate from the decomposition of one and the same PO, i.e., a solvent cage effect is operating (Scheme 1, reactions 3 and 3') [19, 34]. With simple olefins the so-called normal ozonides are over 70 per cent [26]. Cross-ozonide formation is also observed (Scheme 1, reactions 5 and 5'). The amount of cross-ozonides is dependent upon the structure of the double bonds, their concentration in the solution, temperature, and solvent nature. Reference data indicate that the cross-ozonide yield is strongly reduced at C=C bond concentrations of the order of 0.1 M, with non-polar solvents, and at temperatures over 0°C [19]. It is reasonable to expect that the polymeric nature of the double bonds in the polyisoprenes would additionally impede the formation of cross-ozonides. Our estimates showed that the amount of cross-ozonides, formed on ozonolysis of both elastomers, is < 10 per cent of their total quantity. A very small percentage in the overall balance of reacted ozone is the share of the reaction of polymeric ozonides formation (Scheme 1, reactions 4 and 4') [22].

TABLE 10.4 Assignment of the signals in the ¹H-NMR spectra of partially ozonized 1,4-cis-, and 1,4-*trans*-polyisoprenes

Assignment of the signals	Chemical shifts (ppm)		Literature
(according Figure 10.4)	E-IR	Z-IR	
a	Max 1.41	Max 1.40	37, 38
b	1.70–1.78	1.68–1.82	37
		Max 1.71	
c	2.10–2.20	2.10–2.20	33
d	2.22–2.40	2.22–2.35	33
	Max 2.33, 2.36	Max 2.29, 2.33	
e	2.40–2.60	2.35–2.55	33
	Max 2.47	Max 2.48, 2.46	
f	Max 2.73	Max 2.73	33
g	Max 9.76	Max 9.77	33

In the presence of two types of intermediates, DCI and MCI, it is interesting to follow their further conversion with a view to the reaction products determined by ¹H-NMR spectroscopy. The reaction between CI and carbonyl groups is usually considered to be 1,3-cycloaddition [20, 34]. In this connection the CI—aldehyde interaction is the most effective one: it readily precedes with high ozonide yields. Ketones are considerably less dipolarophilic and good yields of ozonides are limited to special conditions involving particularly reactive ketones, intramolecular reactions, or where ketone is used as the reaction solvent and is therefore present in large concentration [19–21]. To evaluate the contribution of reactions 3 and 3' (Scheme 1), it is useful to discuss two hypotheses: (1) no reaction proceeds between MCI and ketone, the registered amount of ozonides being formed in reaction 3; (2) because of the solvent cage effect already mentioned, both the reactions can be considered with practically equal rate constants. Then, the ozonide yields would be proportional to the ratio between the two zwitterions (64:36). In this case, the corresponding yields of the reactions DCI + aldehyde and MCI + ketone can be calculated from the following system of equations:

$$Y = x + y \qquad\qquad (10.6)$$

$$x/y = 64/36 \qquad\qquad (10.7)$$

where Y is the total ozonide yield, x and y are the corresponding ozonide yields with the former and the latter reactions, respectively. Results are given in Table 10.5. A comparison between the aldehyde amounts from both hypotheses and NMR data shows that the former hypothesis leads to results that are much closer to the experimental data.

More complicated is the question of the other conversion routes with carbonyl oxides outside the solvent cage. It is generally accepted that the deactivation of DCI from low molecular olefins normally takes place through a bimolecular reaction mechanism. At low temperatures (below −70°C) dimeric peroxides are formed (Scheme 2, reaction 1), whereas higher temperatures give rise to the formation of carbonyl compounds and evolution of oxygen (Scheme 2, reaction 2) [20]. However, on comparing the yields of keto groups, determined from ¹H-NMR spectra, with the amounts of ketones and DCI, presented in table

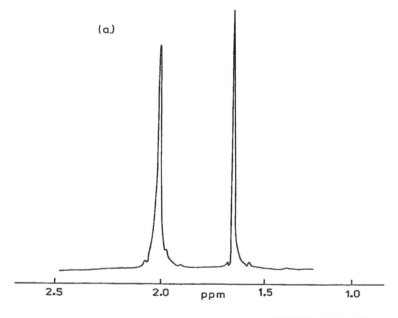

FIGURE 10.6   *(Continued)*

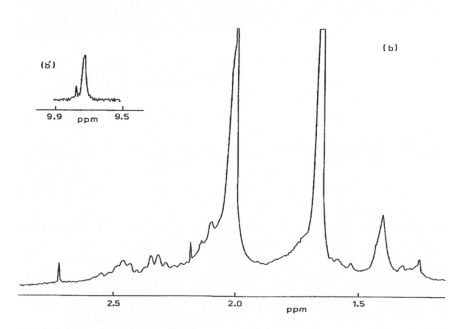

FIGURE 10.6 ¹H-250 MHz NMR spectra of 1,4-*cis*-polyisoprene solutions (1 g/100 ml CCl₄): (a) non-ozonized; (b) ozonized to 23 per cent conversion of the double bonds (external standard TMS; digital resolution 0.4 Hz, 20°C).

5, it is seen that reaction 2, if occurring at all, is not dominating during DCI conversion. From kinetic evaluations of CI concentrations in solution it follows that under comparable experimental conditions these concentrations are four to six orders of magnitude lower than those of the C=C bonds [23]. Very likely, interactions between two or more DCI are hindered by extremely low values of CI concentration and mostly by the polymeric nature of substituent.

R_1. Under these conditions, DCI tautomerization followed by decomposition of hydroperoxides, formed in the excited state (Scheme 2, reactions 3 and 4), seems most probable.

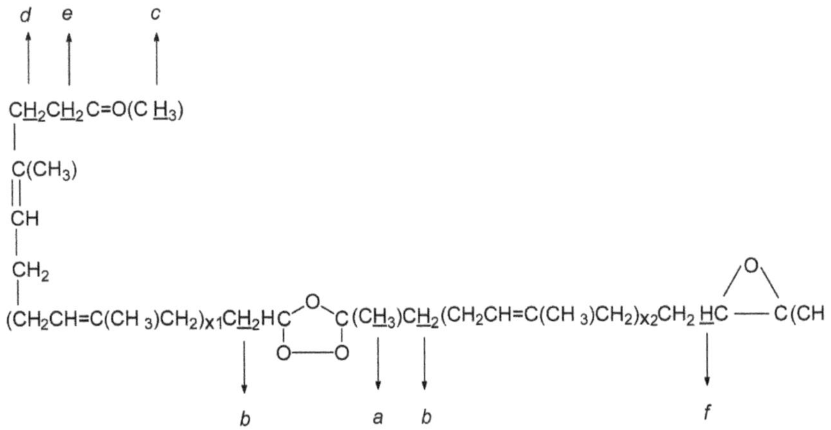

FIGURE 10.7 Selection of protons with characteristic signals in the ^{1}H-250 MHz NMR spectra of partially ozonized polyisoprene macromolecules.

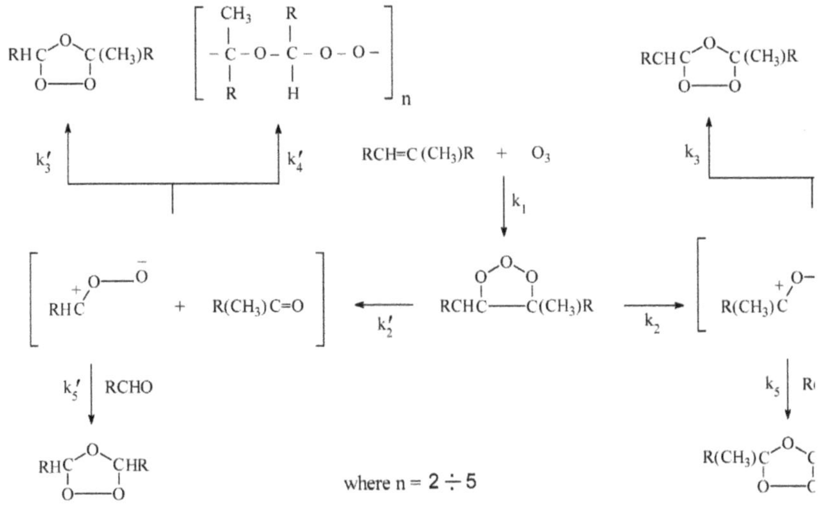

SCHEME 1 Criegee's Mechanism of Formation of Ozonides.

TABLE 10.5 Intermediates and products of the ozonolysis of 1,4-*cis*-, and 1,4-*trans*-polyisoprenes

	DCI + Aldehydes (64%)			MCI + Ketones (36%)		
	Ozonides (%)	Aldehydes (%)	DCI (%)	Ozonides (%)	Ketones (%)	MCI (%)
Hypothese (i)						
E–IR	40	24	24	0	36	36
Z–IR	42	22	22	0	36	36
Hypothese (ii)						
E–IR	26	38	38	14	22	22
Z–IR	27	37	37	15	21	21

On going into MCI conversion routes it appears that MCI has the lower stability and lifetime of these intermediates when compared with DCI. Reference data on the proceeding of reactions 1 and 2 (Scheme 2) during ozonolysis of low-molecular olefins in solution are missing [34]. On ozonolysis of 1,4-*trans*-polychloroprene the share of CI of the MCI type is over 80 per cent, in agreement with the induction effect of the chlorine atom [23]. However, no aldehyde groups in the ozonized polymer solution were found [39]. It follows that the interaction between MCI entities, if any, is not the dominating deactivation reaction with these intermediates. The most probable MCI monomolecular deactivation route is assumed to be the isomerization of the MCI intermediates through hot acid to radicals (Scheme 2, reactions 5, 6, 7), because no acid groups were detected (Scheme 2, reaction 8). According to the literature, reaction 5 (Scheme 2) does not occur with DCI intermediates [21].

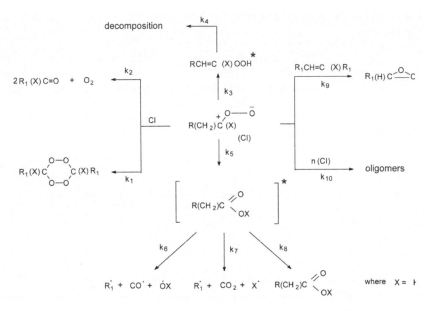

SCHEME 2   Non-Ozonide Routes of Deactivation of Criegee's Intermediates.

Another route of carbonyl oxide deactivation is double-bond epoxidation. Various schemes of olefin epoxidation during ozonolysis have been suggested but the epoxidation through CI (Scheme 2, reaction 9) is presumed to be the most probable with the C=C bonds in polyisoprenes [20, 21]. Since a peak at 2.73 ppm has also been observed under similar conditions on ozonolysis of acrylonitrile—butadiene copolymers [31], as well as during ozonolysis of polybutadienes, it can be assumed that the epoxidation reaction takes place with the participation of both types of CI.

10.4 POLYCHLOROPRENE

The electron-accepting properties of the chlorine atom at the polychloroprene double bond reduces the reactivity of Denka M40 as demonstrated by its relatively low rate constant, i.e., $k = 4 \times 10^3$ M^{-1}s^{-1}. In this reaction the ratio between the zwitterions A and B, according to theoretical calculations, is in favor of A, the ratio being $A/B = 4.55$. The A formation is accompanied by choroanhydryde group formation and that of B with aldehyde one. In both cases the ozonides formation is insignificant and the zwitterions react predominantly in the volume resulting in enhancement of the degradation process. The intensive band detected at 1795 cm^{-1} in the IR spectrum of ozonized Denka M40 solutions (Figure 10.8) is characteristic of chloroanhydride group [39]. This fact correlated well with the conclusion about the direction of the PO decomposition. The band at 955 cm^{-1} is also typical of chloroanhydrides. Two other bands—at 1,044 and 905 cm^{-1} may be attributed to the C–O vibrations. The valent vibrations characteristics of HO-groups are observed in the region of 3,050–3,500 cm^{-1}.

The iodometrical analysis of active oxygen in the ozonized Denka M40 solutions shows that the amount of O–O groups is ~43 per cent. It is of interest to note that the HI reaction with ozonized polychloroprene solutions occurs quantitatively for 3–4 h, while in SKD the same proceeds only to 20 per cent after 24 h. The aforementioned data, however, provide insufficient information for the preferable route of the zwitterions deactivation (via dimerization, polymerization of zwitterions, or secondary processes). The DSC analysis of the products of Denka M40 ozonolyis reveals that the chloroprene rubber ozonolyis yields polyperoxide as the enthalpy of its decomposition is found to be very close to that of dicumeneperoxide (DCP), The higher value of E_a (approximately two times of that of DCP) testifies the possible formation of polymer peroxides [4].

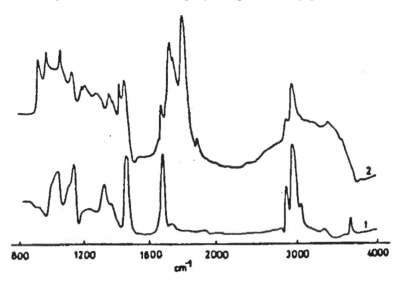

FIGURE 10.8 Infra-red spectra of denka M40 solutions: (1)—non-ozonized; (2)—ozonized to 40 per cent conversion.

10.5   DSC STUDY OF THERMAL DECOMPOSITION OF PARTIALLY OZONIZED DIENE RUBBERS

The thermal decomposition of the ozonized diene rubbers can exercise influence on the ageing processes as an initiator of the oxidation reactions [40]. On the other hand, it is of interest for the elastomers modification and oligomerization [6]. The importance of the DSC method to the investigation of 1,2,4-trioxolanes is based on high values of the enthalpy of the reaction of ozonide thermal decomposition and on the temperature range in which it takes place [41–43]. An intense and relatively broad exothermic peak is characteristic for the thermograms of the partially ozonated 1,4-*cis*-polybutadiene (E-BR), Diene 35 NFA (BR), 1,4-*cis*-polyisoprene (E-IR), 1,4-*trans*-polyisoprene (Z-IR), and 1,4-*trans*-polychloroprene (PCh), recorded in the range of 60–200°C (Figure 10.9). In practical, no thermal effects are detected in the respective thermograms of the nonozonized samples.

For an objective consideration on the enthalpy changes it is necessary to normalize ΔH values with respect to the amount of consumed ozone (G). It is distributed on the whole mass of the ozonated sample. The whole mass is a sum of the mass of nonozonated rubber and mass of the incorporated in diene macromolecules oxygen atoms, grouped as a moles ozone. Determination of the amounts of the incorporated ozone (G_{inc}) in rubber samples is performed in [43]. The corresponding values of the coefficient of incorporation of ozone (c_{inc}), defined as a ratio of the amounts of incorporated (G_{inc}) and consumed (G) ozone is Presented in Table 10.6. Using ΔH_1 values and ratio of the functional group, deduced from the NMR spectra, the enthalpy of the ozonide thermal decomposition (ΔH_2) has been evaluated. The data of Table 10.6 clearly show tendency of increasing of ΔH_2 values with the number of alkyl substituents of the ozonides.

There are no considerable differences between the values of the activation energy (E) and reaction order (n) of the ozonide thermal decomposition with E-IR and Z-IR (Table 10.7). The smaller E values of the polyisoprene ozonides in comparison with those of E-BR ozonides and 1-decene ozonide are, most probably, due to the lower thermal stability of small amounts of oligomeric peroxides, which are present among the reaction products of E-IR and Z-IR ozonolysis [36].

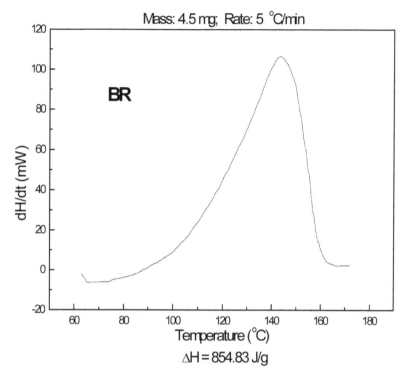

FIGURE 10.9   DSC curve of partially ozonated diene 35 NFA rubber (BR).

TABLE 10.6   Enthalpy changes on the thermolysis of partially ozonated diene rubbers

Sample	ΔH (J/g)	c_{inc}	ΔH_1, (kJ/mole O_3)	Ozonide (mole)/ incorp. Ozone (mole)	ΔH_2 (kJ/ mole ozon- ide)
E-BR	954.07	0.93	332	0.96	373
BR	854.83	0.87	271	0.89	350
E-IR	576.19	0.62	182	0.65	453
Z-IR	576.62	0.62	178	0.67	426
PCh	704.57	0.81	254		
1-Decene Ozonide					349 [Ref. 7]

TABLE 10.7   DSC analysis of the ozonized rubbers decomposition [43]

	$*T_m$ (°C)	α_s	E (kJ/mole)	N
E-BR	147	0.67	125	1.0
BR	145	0.68	117	0.96
E-IR	160	0.68	111	1.04
Z-IR	149	0.68	113	1.08
PCh	96	0.55	70	1.25
1-Decene ozonide	117	0.63	132	1.0

$*T_m$ and α_s Values at Heating rate of 5°C/min; T_m and α_s are Temperature and Degree of Conversion at the DSC peak Maximum, respectively.

10.6   CONCLUSIONS

The ozone reaction with a number of polyidienes with different configurations of the double bonds and various substituents was investigated in CCl_4 solution. The changes of the viscosity of the polymer solutions during the ozonolysis were characterized by the determination of the number of chain scissions per molecule of reacted ozone (φ). The influence of the conditions of mass-transfer of the reagents in a bubble reactor on the respective φ values was discussed.

The basic functional groups—products from the rubbers ozonolysis were identified and quantitatively characterized by means of IR-spectroscopy and ^{1}H-NMR spectroscopy. The aldehyde—ozonide ratio was 11:89 and 27:73 for E-BR and BR, respectively. In addition, epoxide groups were detected, only in the case of BR, their yield was about 10 per cent of that of the aldehydes. On polyisoprenes the ozonide—ketone—aldehyde ratio was 40:37:23 and 42:39:19 for E-IR and Z-IR, respectively. Besides the already-specified functional groups, epoxide groups were also detected, their yields being 8 and 7 per cent for E-IR and Z-IR, respectively, with respect to reacted ozone. In the case of 1,4-*trans*-polychloroprene, the chloroanhydride group was found to be the basic carbonyl product.

A reaction mechanism that explains the formation of all identified functional groups was proposed. It has been shown that the basic route of the reaction of ozone with elastomer double bonds—the formation of normal ozonides does not lead directly to a decrease in the molecular mass of the elastomer macromolecules, because the respective 1,2,4-trioxolanes are relatively stable at ambient temperature. The most favorable conditions for ozone degradation emerge when the cage interaction between Crigee's intermediates and respective carbonyl groups does not proceed. The amounts of measured different carbonyl groups have been used as an alternative way for evaluation of the intensity and efficiency of the ozone degradation.

The thermal decomposition of partially ozonated diene rubbers was investigated by DSC. The respective values of the enthalpy, the activation energy, and the reaction order of the 1,2,4-trioxolanes were determined.

KEYWORDS

- ^{1}H-NMR spectroscopy
- Monomers
- Ozonation
- Ozonolysis
- Polyisobutylene

REFERENCES

Zuev, Yu. S.; Degradation of Polymers by Action of Aggressive Medias. Moscow, Russia: Khimia Publishing House; **1972**.

Razumovskii, S. D.; and Zaikov, G. E.; In: Devepments in Polymer Stabilisation. Ed. Scott, G.; London: Applied Science Publishing House; **1983**, *6*.

Jellinek, H. H. G.; Aspects of Degradation and Stabilisation of Polymers. Ed. Jellinek, H. H. G.; Amsterdam, Nederland: Elsevier; **1978**, Chapter 9.

Zaikov, G. E.; and Rakovsky, S. K.; Ozonation of Organic & Polymer Compounds. Smithres Rapra, UK; **2009**, Chapters 3 and 4.

Sweenew, G. P.; Ozonolysis of natural rubber. *J. Polym. Sci. Part A-1*, **1968**, *6, 2679*.

Egorova, G. G.; and Shagov, V. S.; "Ozonolysis in the chemistry of unsaturated polymers". In: Synthesis and Chemical Transformation of Polymers. Izd. LGU, Leningrad; **1986**, (in Russian).

Hackathorn, M. J.; and Brock, M. J.; The determination of "Head-head" and "Tail-tail" structure in polyisoprene. *Rubb. Chem. Technol.* **1972,** *45(6), 1295.*

Hackathorn, M. J.; and Brock, M. J.; Altering structures in copolymers as elucidated by microozonolysis. *J. Polym. Sci. Part A-1,* **1968,** *6, 945.*

Beresnev, V. V.; and Grigoriev, E. I.; Infuence of the ozonation conditions on molecular mass distribution of oligodienes with end functional groups. *Kautchuk i Rezina.* **1993,** *4,* 10.

Nor, H. M.; and Ebdon, J. R.; Telechelic, liquid natural rubber: a review. *Prog. Polym. Sci.* **1998,** *23,* 143.

Solanky, S. S.; and Singh, R. P.; Ozonolysis of natural rubber: a critical review. *Prog. Rubb. Plastics. Technol.* **2001,** *17(1), 13.*

Robin, J. J.; The use of ozone in the synthesis of new polymers and the modification of polymers. *Adv. Polym. Sci.* **2004,** *167, 35.*

Murray, R.; and Story, P.; Ozonation. In: Chemical Reactions of Polymers. Chapter 3, Mir, Moscow; **1967.**

Cataldo, F.; The action of ozone on polymers having unconjugated and cross-or linearly conjugated unsaturation: chemistry and technological aspects. *Polym. Deg. Stab.* **2001,** *73, 511.*

Montaudo, G.; Scamporrino, E.; Vitalini, D.; and Rapisardi, R.; Fast atom bombardment mass spectrometric analysis of the partial ozonolysis products of poly(isoprene) and popy(chloroprene). *J. Polym. Sci.:* Part A, **1992,** *30, 525.*

Ivan, G.; and Giurginca, M.; Ozone destruction of some trans-polydienes. *Polym. Deg. Stab.* **1998,** *62, 441.*

Nor, H. M.; and Ebdon, J. R.; Ozonolisys of natural rubber in chloroform solution. Patr 1 A study by GPC and FTIR spectroscopy. *Polym.* **2000,** *41, 2359.*

Somers, A. E.; Bastow, T. J.; Burgar, M. I.; Forsyth, M.; and Hill, A. J.; Quantifying rubber degradation using NMR. *Polym. Deg. Stab.* **2000,** *70, 31.*

Bailey, P. S.; *Ozonation in organic chemistry.* New York: Academic Press; **1978, 1982,** *1, 2.*

Bunnelle, W. H.; Preparation, properties, and reactions of carbonyl oxides. *Chem. Rev.* **1991,** *91, 335.*

McCullough, K. J.; and Nojima, M.; "Peroxides from ozonization". In: Organic Peroxides. Ando, W.; Chapter 13, Ed., John Wiley & Sons Ltd.; **1992.**

Anachkov, M. P.; Rakovsky, S. K.; Shopov, D. M.; Razumovskii, S. D.; Kefely, A. A.; and Zaikov, G. E.; Study of the ozone degradation of polybutadiene, polyisoprene and polychloroprene in solution. *Polym. Deg. Stab.* **1985,** *10(1), 25.*

Razumovskii, S. D.; Rakovski, S. K.; Shopov, D. M.; Zaikov, G. E.; Ozone and its Reactions with Organic Compounds (in Russian). Sofia: Publication House of Bulgarian Academy of Sciences; **1983.**

Flory, P. J.; Principles of Polymer Chemistry. New York; **1953,** 41 p.

Rakovsky, S.; and Zaikov, G.; Ozonolysis of polydienes. *J. Appl. Polym. Sci.* **2004,** *91(3), 2048.*

Berlin, A. A.; Sayadyan, A. A.; Enikolopyan, N. S.; Molecular-weight distribution and reactivity of polymers in concentrated and diluted solutions. *Visoko Molekul. Soedin-*A, **1969,** *11, 1893.*

Razumovskii, S. D.; Niazashvili, G. A.; Yur'ev, Yu. N.; Tutorskii, I. A.; Synthesis and study of polymeric ozonides. *Visoko Molekul. Soedin.* A, **1971,** *13, 195.*

Razumovskii, S. D.; and Zaikov, G. E.; Kinetics and mechanism of ozone reaction with double bonds. *Uspekhi Khimii.* **1980,** *49, 2344.*

Kefely, A. A.; Razumovskii, S. D.; and Zaikov, G. E.; Reaction of ozone with polyethylene. *Visoko Molekul Soedin.* A, **1971**, *13, 803.*

Rakovsky, S. K.; Cherneva, D. R.; Shopov, D. M.; and Razumovskii, S. D.; Applying of barbotage method to investigation the kinetic of ozone reactions with organic compounds. *Izv. Khim. BAN.* **1976,** *XI(4), 711.*M.P. Anachkov, S.K. Rakovsky, R.V. Stefanova and D.M. Shopov, Kinetics and mechanism of the ozone degradation of nitrile rubbers in solution, *Polym. Deg. Stab.,* <u>19</u>*(2), 293* (**1987**).

Anachkov, M. P.; Rakovsky, S. K.; and Zaikov, G. E.; Ozonolysis of polybutadienes with different microstructure in solution. *J. Appl. Polym. Sci.* **2007,** *104(1), 427.*

The Sadtler Handbook of Proton NMR Spectra. Philadelphia, PA: Sadtler; **1978.**

Kuczkowski, R. L.; The structure and mechanism of formation of ozonides. *Chem. Soc. Rev.* **1992,** *21, 79.*

Nakanishi, K.; Infrared Spectra and the Structure of Organic Compounds. Moscow: Mir; **1965.**

Anachkov, M. P.; Rakovsky, S. K.; and Stefanova, R. V.; Ozonolysis of 1,4-cis-polyisoprene and 1,4-trans-polyisoprene in solution. *Polym. Deg. Stab.* **2000,** *67(2), 355.*

Murray, R. W.; Kong, W.; and Rajadhyaksha, S. N.; The ozonolysis of tetra-methylethylene - concentration and temperature effects. *J. Org. Chem.* **1993,** *58, 315.*

Choe, J. I.; Sprinivasan, M.; and Kuczkowski, R. L.; Mechanism of the ozonolysis of propene in the liquid-phase. *J. Am. Chem. Soc.* **1983,** *105, 4703.*

Anachkov, M. P.; Rakovsky, S. K.; Stefanova, R. V.; and Stoyanov, A. K.; Ozone degradation of polychloroprene rubber in solution. *Polym. Deg. Stab.* **1993,** *41(2), 185.*

Fotty, R. K.; Rakovsky, S. K.; Anachkov, M. P.; Ivanov, S. K.; Cumene oxidation in the presence of 1-decene-ozonide. *Oxidat. Commun.* **1997,** *20(3), 411.*

Anachkov, M. P.; Rakovsky, S. K.; Stoyanov, A. K.; and Fotty, R. K.; DSC study of the thermal decomposition of 1-decene ozonide. *Thermochim. Acta.* **1994,** *237, 213.*

Anachkov, M. P.; Rakovsky, S. K.; and Stoyanov, A. K.; DSC study of the thermal decomposition of partially ozonized diene rubbers. *J. Appl. Polym. Sci.* **1996,** *61, 585.*

Anachkov, M. P.; Rakovsky, S. K.; Enthalphy of the thermal decomposition of functional groups of peroxide type. Formed by Ozonolysis of diene rubbers in solution. *Bulg. Chem. Comm.* **2002,** *34(3/4), 486.*

INDEX